Advanced Statistical Analysis for Soil Scientists

Tancredo Souza

Advanced Statistical Analysis for Soil Scientists

Springer

Tancredo Souza
Department of Soil and Rural Engineering
Federal University Paraiba
Areia, Paraíba, Brazil

ISBN 978-3-031-88160-2 ISBN 978-3-031-88161-9 (eBook)
https://doi.org/10.1007/978-3-031-88161-9

This Springer imprint is published by the registered company Springer Nature Switzerland AG
The registered company address is: Gewerbestrasse 11, 6330 Cham, Switzerland

To my wife, Fransuene Dias
She supported me in ways no one else could during my darkest days.
To my beloved mother, Rejane Souza
She was truly a force of nature. Her life was a blessing, her memory a treasure, and her love and dedication helped me grow into the professional I am today.
"The soul is placed in the body like a rough diamond, and must be polished, or the luster of it will never appear."—Daniel Defoe.

Preface

In the dynamic and multifaceted field of soil science, the application of advanced statistical analysis has become indispensable for young scientists striving to unveil the intricate relationships and patterns inherent within soil ecosystems. This handbook, "Advanced Statistical Analysis for Soil Scientists," aims to bridge the gap between traditional soil analysis methods and contemporary statistical techniques, empowering emerging researchers with the tools necessary to conduct robust and insightful investigations. By incorporating methods such as PERMANOVA, ANOSIM, PCA, NMDS, RDA, and SEM, we endeavor to provide a comprehensive guide that underscores the critical role of these advanced analyses in elucidating complex soil processes and interactions.

Understanding soil ecosystems requires more than just empirical observations; it necessitates a rigorous analytical approach that can discern subtle variations and multidimensional relationships. Techniques like Principal Component Analysis (PCA) and Non-Metric Multidimensional Scaling (NMDS) allow scientists to reduce the dimensionality of complex data sets, making it possible to visualize and interpret underlying patterns. Redundancy Analysis (RDA) and Structural Equation Modeling (SEM) further enhance our ability to model and quantify the direct and indirect interactions within soil systems. This handbook illustrates these methodologies through practical examples and case studies, demonstrating their application in real-world soil science research and their potential to yield meaningful insights.

For young scientists, mastering these advanced statistical techniques is crucial for confidently reporting their findings and contributing to the broader scientific community. This handbook not only provides a theoretical foundation but also emphasizes practical implementation, offering step-by-step guidance on data preparation, analysis, and interpretation. By equipping the next generation of soil scientists with these essential skills, we aim to foster a deeper understanding of soil ecosystems and promote the advancement of sustainable land management practices.

Bananeiras, Paraiba, Brazil
January 10, 2025

Tancredo Souza

Contents

About the Author

Tancredo Souza This book was written while Tancredo Souza was a Visiting Professor at the Department of Agriculture, Federal University of Paraiba, Bananeiras, Brazil. His research focuses on the effects of land use on the diversity of soil organisms in tropical and subtropical ecosystems. He is particularly interested in plant–soil interactions, the ecology of mycorrhizal fungi, and biogeochemistry. Additionally, he has expertise in multivariate statistical analysis, applying techniques such as PCA, CCA, RDA, NMDS, and SEM to interpret complex soil data. He holds a Bachelor's degree in Agronomy, a Master's degree in Soil and Water Management and Conservation, and a Ph.D. in Soil Sciences from the Federal University of Paraíba, Brazil, and the University of Coimbra, Portugal.

PERMANOVA

1

Abstract

PERMANOVA (Permutational Multivariate Analysis of Variance) differs from traditional ANOVA by being a non-parametric test that operates on distance matrices rather than raw data, making it suitable for complex, multivariate datasets without the need for normal distribution or homogeneity of variances. The primary assumptions for PERMANOVA include the independence of observations, appropriate choice of distance measures, and the need to check for homogeneity of group dispersions to avoid misleading results. PERMANOVA effectively handles high-dimensional ecological data by utilizing distance matrices and permutation techniques, allowing for the analysis of multiple interrelated variables simultaneously. This flexibility and robustness make PERMANOVA particularly useful in soil science research, where data often exhibit high variability and complexity, such as in studies of soil microbial communities, physical properties, and organic carbon pools. Before applying PERMANOVA, researchers should ensure the assumptions are met, particularly checking for homogeneity of dispersions using methods like PERMDISP, to ensure reliable and interpretable results.

Keywords

Distance measures · Multivariate data · Non-parametric analysis · PERMANOVA · Soil biodiversity

Questions Covered in the Chapter

(a) What is the primary difference between PERMANOVA and traditional ANOVA?
(b) What are the main assumptions required for PERMANOVA?
(c) How does PERMANOVA handle high-dimensional ecological data?

T. Souza, *Advanced Statistical Analysis for Soil Scientists*,
https://doi.org/10.1007/978-3-031-88161-9_1

(d) Why is PERMANOVA particularly useful in soil science research?
(e) What should researchers check for before applying PERMANOVA to ensure reliable results?

Introduction

PERMANOVA, or Permutational Multivariate Analysis of Variance, is a robust non-parametric statistical test used to compare the differences between groups of multivariate data (Anderson 2017). Unlike traditional ANOVA, which relies on parametric assumptions, PERMANOVA utilizes permutation techniques to assess the significance of the observed differences. This method is particularly valuable in ecological and environmental studies, where data often deviate from normality and homogeneity of variances (Botha et al. 2024). By relying on distance matrices rather than raw data, PERMANOVA provides a flexible and powerful approach to analyze complex ecological datasets.

The primary assumptions of PERMANOVA include the independence of observations and the appropriate choice of a distance measure that accurately reflects the ecological distances between samples. While PERMANOVA does not require the assumption of multivariate normality, it is sensitive to differences in group dispersions, which can affect the interpretation of results. To mitigate this, researchers should carefully check for homogeneity of dispersions using methods such as PERMDISP before applying PERMANOVA. Ensuring these assumptions are met is crucial for obtaining reliable and interpretable results.

One of the significant advantages of PERMANOVA as an advanced statistical analysis is its ability to handle complex, high-dimensional data without the stringent assumptions required by parametric tests. This flexibility makes it ideal for analyzing ecological data, where variables are often interdependent and non-normally distributed. Additionally, PERMANOVA can incorporate various distance measures, such as Bray-Curtis or Euclidean distances, allowing researchers to tailor the analysis to their specific data characteristics. Its non-parametric nature also makes PERMANOVA robust against outliers and other data irregularities, enhancing its applicability in real-world research scenarios.

In the field of soil science, PERMANOVA is particularly important for studying the diversity and composition of soil organisms under different land use practices and environmental conditions (Canini et al. 2024). By enabling the comparison of complex microbial communities and soil properties across various treatments, PERMANOVA helps researchers understand the impact of anthropogenic activities on soil health and ecosystem functioning. Its application can reveal critical insights into soil biodiversity, resilience, and the interactions between soil organisms and their environment, thereby informing sustainable land management and conservation strategies (Dietrich et al. 2024).

Primary Difference Between PERMANOVA and Traditional ANOVA

The primary difference between PERMANOVA (Permutational Multivariate Analysis of Variance) and traditional ANOVA (Analysis of Variance) lies in their underlying statistical assumptions and the types of data they are designed to handle. Traditional ANOVA is a parametric test that compares the means of different groups based on the assumption that the data are normally distributed and have homogenous variances (homoscedasticity). It relies on calculating the variance within and between groups to determine if there are significant differences among the group means.

PERMANOVA, on the other hand, is a non-parametric test that does not assume normal distribution or homogeneity of variances. Instead of working directly with the raw data, PERMANOVA operates on distance matrices, which quantify the dissimilarities between pairs of samples. By using permutation techniques to assess the significance of observed differences, PERMANOVA can analyze complex, multivariate data more flexibly. This makes it particularly suitable for ecological and environmental studies, where data often do not meet the strict assumptions required by traditional ANOVA.

For example, in soil science, researchers might want to compare the microbial communities in soils under different land use practices, such as agricultural fields, forests, and urban areas. Traditional ANOVA would require the microbial abundance data to be normally distributed and exhibit homogenous variances across the groups, which is often not the case due to the high variability and complexity of ecological data. PERMANOVA, however, can handle these complexities by constructing a distance matrix (e.g., using Bray-Curtis dissimilarity) to measure the differences in microbial community composition between the samples. It then uses permutations to test whether the observed differences are statistically significant (Forsmark et al. 2024).

Furthermore, PERMANOVA's reliance on distance matrices allows it to incorporate various distance measures tailored to specific data characteristics. For instance, in soil science, researchers might use Euclidean distance for analyzing soil chemical properties or Bray-Curtis dissimilarity for microbial community data. This flexibility enables a more accurate and meaningful analysis of multivariate ecological data. Traditional ANOVA lacks this adaptability, making it less suitable for handling the diverse and often non-normal data encountered in soil science research. Thus, PERMANOVA offers a powerful alternative for ecologists and soil scientists aiming to uncover significant patterns and relationships in complex, real-world datasets (Jaeger et al. 2024).

Key Concepts of PERMANOVA

PERMANOVA operates on a distance matrix, which quantifies the dissimilarity between pairs of samples (Anderson and Walsh 2013). This is crucial in soil science, where data can be complex and multidimensional. For instance, when studying soil chemical properties across different land uses, researchers might measure a range of variables like pH, nutrient levels, and organic content. Traditional parametric tests might not handle these variables well due to their non-normal distributions. By using distance measures such as Euclidean distance for soil chemical properties, PERMANOVA can effectively capture and compare the overall chemical profile of soils from different sites, providing a comprehensive analysis of soil health and fertility.

The use of permutations is another fundamental aspect of PERMANOVA. Instead of relying on the parametric assumptions required by traditional ANOVA, PERMANOVA assesses significance by repeatedly permuting the data and recalculating the test statistic. This process builds a distribution of the statistic under the null hypothesis, allowing for a robust significance test even with non-normally distributed data. For example, in examining microbial communities in tropical soils, researchers can use Bray-Curtis dissimilarity to construct a distance matrix. By permuting the data, they can determine whether observed differences in microbial composition between forested and agricultural soils are statistically significant, providing insights into how land use impacts soil microbiota (Jiang et al. 2024).

The pseudo-F statistic in PERMANOVA serves a role similar to the F-statistic in traditional ANOVA, quantifying the ratio of between-group variance to within-group variance based on the chosen distance measure. This allows researchers to understand the extent of variation attributable to different treatments or conditions. For example, when analyzing the chemodiversity of soil organic matter in tropical soils, researchers might use Jaccard distance to capture the diversity of organic compounds present. The pseudo-F statistic helps determine whether differences in chemodiversity between natural and managed ecosystems are significant, aiding in the assessment of how management practices affect soil organic matter composition (Kelly et al. 2015).

Finally, the null hypothesis in PERMANOVA is that the centroids (multivariate means) of the groups are the same in the space defined by the distance measure. This means that any observed differences are due to random variation rather than a true effect (Anderson and Walsh 2013). By testing this hypothesis, researchers can confidently report whether land use, environmental conditions, or management practices lead to significant changes in soil properties, microbial communities, or organic matter chemodiversity. This capability is essential for advancing our understanding of soil ecosystems and developing strategies for sustainable soil management in tropical regions.

To Summarize

Distance Matrix: PERMANOVA works on a distance matrix, which quantifies the dissimilarity between pairs of samples. Common distance measures include Euclidean distance, Bray-Curtis dissimilarity, and Jaccard distance.

Permutations: Instead of relying on parametric assumptions, PERMANOVA assesses significance by repeatedly permuting the data and recalculating the test statistic to build a distribution of the statistic under the null hypothesis.

Pseudo-F Statistic: Analogous to the F-statistic in traditional ANOVA, the pseudo-F statistic in PERMANOVA quantifies the ratio of between-group variance to within-group variance based on the chosen distance measure.

Null Hypothesis: The null hypothesis in PERMANOVA is that the centroids (multivariate means) of the groups are the same in the space defined by the distance measure.

Main Assumptions Required for PERMANOVA

PERMANOVA (Permutational Multivariate Analysis of Variance) is a flexible and powerful statistical tool, but it still relies on certain assumptions to ensure valid and interpretable results. One of the primary assumptions of PERMANOVA is the independence of observations. This means that the data points in the dataset should not be influenced by each other. In soil science, for example, when comparing soil properties of Regosols and Ferralsols, samples should be collected in a way that ensures each sample is independent. This can be achieved by sampling from different locations or ensuring sufficient spatial separation to avoid spatial autocorrelation (Anderson 2017).

Another critical assumption of PERMANOVA is the appropriate choice of a distance measure that accurately reflects the ecological or biological distances between samples. The distance measure should be chosen based on the type of data and the research question. For example, in soil science, Euclidean distance might be suitable for comparing soil chemical properties, whereas Bray-Curtis dissimilarity could be more appropriate for comparing microbial community compositions. When studying the differences between Regosols and Ferralsols, researchers must select a distance measure that accurately captures the dissimilarities in soil properties or microbial communities to ensure meaningful comparisons.

While PERMANOVA does not require multivariate normality, it is sensitive to differences in group dispersions, which can affect the interpretation of results. This means that the spread or variability within each group should be similar. If one group has much larger variability than another, it can lead to misleading conclusions. To address this, researchers should check for homogeneity of dispersions using methods like PERMDISP (Permutational Analysis of Multivariate Dispersions) before applying PERMANOVA. For instance, when comparing soil properties of Regosols and Ferralsols, it is essential to ensure that the variability within each soil type is similar. If Regosols exhibit much higher variability in pH or nutrient levels than Ferralsols, this disparity should be accounted for to avoid biased results.

Finally, PERMANOVA assumes that the chosen distance measure appropriately captures the ecological distances between samples. This means that the distance measure should reflect the actual differences in the ecological or biological attributes being studied. In soil science, for example, when comparing the microbial communities of agroecological and conventional farming systems, the chosen distance measure (e.g., Bray-Curtis dissimilarity) should accurately represent the differences in microbial compositions. If the distance measure does not appropriately reflect these differences, the results of the PERMANOVA may not be meaningful or accurate. Ensuring that the distance measure is suitable for the data and research question is crucial for the validity of the analysis.

Steps to Conduct PERMANOVA

The first step in conducting PERMANOVA is to define the research question clearly. This involves specifying the hypothesis you want to test, such as whether there are significant differences in soil microbial communities under different land-use types. For example, you might hypothesize that the bacterial community structure in soils varies between agricultural fields and forested areas. Defining a clear research question helps focus the study and ensures that the subsequent steps are aligned with the objectives.

Next, you need to collect multivariate data relevant to your research question. In soil science, this could include a wide range of measurements. For instance, if you are studying soil physical properties, you might measure variables such as soil texture, bulk density, and porosity. If your focus is on bacterial communities, you would collect data on species composition and abundance. Similarly, for studies on soil organic carbon pools in organic soils, you would gather data on different carbon fractions, such as particulate organic carbon, mineral-associated organic carbon, and dissolved organic carbon. Collecting high-quality, comprehensive data is crucial for the success of the analysis (Le Roux et al. 2024).

After data collection, the next step is to choose an appropriate distance measure based on your data type and research question. The distance measure quantifies the dissimilarity between pairs of samples. For example, Bray-Curtis dissimilarity is commonly used for ecological community data, making it suitable for comparing bacterial community structures in different soil samples. Euclidean distance might be more appropriate for soil physical properties, as it measures straight-line distances between points in multivariate space. For soil organic carbon pools, a distance measure that captures the differences in carbon fractions, such as Jaccard distance, might be used. Selecting the right distance measure ensures that the analysis accurately reflects the underlying ecological or biological differences (Lejoly et al. 2024).

The final steps involve creating a distance matrix, performing PERMANOVA, and interpreting the results. Once you have chosen the distance measure, compute the distance matrix from your multivariate data. This matrix forms the basis for the PERMANOVA test. Using software like R (with the vegan package), PRIMER, or

PAST, conduct the PERMANOVA test by specifying the distance matrix and the grouping factor (e.g., land-use type). The software will calculate the pseudo-F statistic and associated p-value, which indicate whether there are significant differences between the groups. For example, if studying soil organic carbon pools, the results might show significant differences in carbon fractions between organic and conventional farming systems. Interpreting these results helps determine the impact of different factors on soil properties and provides insights into soil health and management practices.

To Summarize

Define the Research Question: Specify the hypothesis you want to test, such as differences in soil microbial communities under different land-use types.

Collect Data: Gather multivariate data. In soil science, this might include measurements of soil properties (e.g., pH, nutrient levels) or soil microbial communities (e.g., species composition).

Choose a Distance Measure: Select an appropriate distance measure based on your data type and research question. For example, Bray-Curtis dissimilarity is commonly used for ecological community data.

Create a Distance Matrix: Compute the distance matrix from your multivariate data.

Perform PERMANOVA: Use software to conduct the PERMANOVA test. The test involves specifying the distance matrix and the grouping factor (e.g., land-use type).

Interpret Results: Assess the pseudo-F statistic and the associated p-value to determine if there are significant differences between the groups.

Handling High-Dimensional Ecological Data with PERMANOVA

PERMANOVA is particularly adept at handling high-dimensional ecological data, which is common in fields like soil science. High-dimensional data refers to datasets with a large number of variables or measurements, often collected from complex ecological systems. PERMANOVA handles such data by using distance matrices to quantify the dissimilarities between samples based on all measured variables simultaneously, rather than reducing the data to a few summary statistics.

In the context of soil science, consider a study of Agroforestry systems, which integrate trees with crops or livestock. Researchers might measure various soil properties such as nutrient levels (nitrogen, phosphorus, potassium), soil organic carbon, microbial biomass, and soil moisture. Instead of analyzing each variable separately, PERMANOVA can assess the collective impact of these variables, providing a holistic view of soil health and ecosystem function. One-way PERMANOVA manages high-dimensional data is by constructing a distance matrix that represents the multivariate differences between samples. For instance, in an Agroforestry system, you could use the Bray-Curtis dissimilarity measure to capture the differences in soil microbial community composition between areas under Agroforestry and conventional monoculture systems. The distance matrix encompasses all the measured microbial taxa, allowing PERMANOVA to detect overall community changes rather than focusing on individual species (Newman et al. 2024).

The use of permutation techniques is another strength of PERMANOVA when dealing with high-dimensional data. Traditional parametric tests often struggle with the complex distributions and interdependencies of high-dimensional ecological data. PERMANOVA, however, relies on permuting the data to build a null distribution of the test statistic. This approach is robust against the non-normality and heteroscedasticity that often characterize ecological data. For example, in evaluating the impact of Agroforestry on soil nutrient dynamics, PERMANOVA can handle the interrelated changes in multiple soil nutrients, providing a comprehensive assessment of how Agroforestry practices influence soil nutrient profiles compared to monocultures (Peña-Gómez et al. 2024).

Finally, PERMANOVA's flexibility in choosing distance measures allows it to tailor the analysis to the specific characteristics of high-dimensional data. In the case of Agroforestry systems, researchers might use Euclidean distance for continuous soil properties like pH and nutrient levels, and Bray-Curtis dissimilarity for categorical data like microbial community composition. This adaptability ensures that the analysis accurately captures the ecological complexities of the system, making PERMANOVA a powerful tool for high-dimensional ecological studies (Segura et al. 2024).

Why PERMANOVA Is Particularly Useful in Soil Science Research?

PERMANOVA (Permutational Multivariate Analysis of Variance) is particularly useful in soil science research due to its ability to handle complex, high-dimensional, and non-normally distributed data. This flexibility is crucial for analyzing the intricate interactions and diverse variables present in soil ecosystems, especially in unique and heterogeneous environments like Amazonian dark earth (Terra Preta).

(a) Handling High-Dimensional Data: Soil science research often involves numerous variables, such as chemical properties (pH, nutrient levels), physical properties (texture, moisture content), and biological properties (microbial communities). Traditional statistical methods may struggle with the high dimensionality and interdependencies of these variables. PERMANOVA, however, can analyze all these variables simultaneously by constructing a distance matrix that represents the multivariate differences between samples. For example, when studying Amazonian dark earth, researchers can assess the combined impact of its rich nutrient content, unique microbial communities, and altered soil structure, providing a comprehensive analysis of its properties and functions (Peña-Gómez et al. 2024).

(b) Non-Parametric Nature: One of PERMANOVA's key strengths is its non-parametric nature, meaning it does not rely on the assumption of normal distribution or homogeneity of variances. Soil data often deviate from these assumptions due to natural variability and complex interactions. PERMANOVA uses permutation techniques to assess the significance of observed differences, making it robust against such deviations. For instance, in examining the microbial diversity of Amazonian dark earth compared to surrounding soils, PERMANOVA can effectively handle the non-normal distribution of microbial community data, providing reliable results without the stringent requirements of parametric tests (Shulman et al. 2024).
(c) Flexibility in Distance Measures: PERMANOVA's flexibility in choosing distance measures allows researchers to tailor the analysis to the specific characteristics of their data. Different distance measures can be used depending on the nature of the variables. For example, Bray-Curtis dissimilarity can be used to compare microbial community compositions in Amazonian dark earth, while Euclidean distance might be suitable for analyzing variations in soil chemical properties. This adaptability ensures that the analysis accurately reflects the ecological and biological differences relevant to the research question (Peña-Gómez et al. 2024).

Practical Applications in Amazonian Dark Earth Studies

Amazonian dark earth is known for its high fertility and unique composition, largely due to human activity. Researchers might use PERMANOVA to compare the soil properties and microbial communities of Amazonian dark earth to adjacent non-anthropogenic soils. By doing so, they can assess the impact of historical land management practices on soil health and biodiversity. For example, a study might measure soil organic carbon, nutrient levels, and microbial diversity across different soil types. Using PERMANOVA, researchers can determine if the differences observed are statistically significant and linked to the anthropogenic origins of Amazonian dark earth, providing insights into sustainable soil management practices. PERMANOVA's ability to handle high-dimensional, non-normally distributed data, along with its flexibility in distance measures, makes it an invaluable tool in soil science research. Its application in studies of Amazonian dark earth exemplifies how it can provide deep insights into complex soil ecosystems, enhancing our understanding and informing sustainable land management practices (Peña-Gómez et al. 2024).

Applying PERMANOVA in Soil Science

PERMANOVA is an invaluable tool in soil science for analyzing complex, multivariate data. This statistical method is particularly useful when dealing with multiple variables that interact in ways that traditional parametric tests cannot easily handle. One of the key applications of PERMANOVA in soil science is in the study of soil microbial communities. For example, researchers can use PERMANOVA to compare bacterial community compositions across different land-use types or management practices. By analyzing the entire community structure, rather than individual species, PERMANOVA can reveal significant differences in microbial composition between soils under organic farming, conventional farming, and natural ecosystems, providing insights into how different practices affect soil health (Sun et al. 2024; Tang et al. 2016).

Another important application of PERMANOVA is in assessing differences in soil physical properties across various sites or treatments. Soil physical properties, such as texture, bulk density, and moisture content, can vary widely depending on land management practices and environmental conditions. For instance, in a study comparing the physical properties of soils in Agroforestry systems versus monoculture systems, PERMANOVA can be used to analyze the combined effects of these properties. This comprehensive analysis helps to identify significant differences in soil structure and compaction, which are crucial for understanding soil fertility and plant growth (Wang et al. 2024).

Environmental impact studies are another area where PERMANOVA shines. Researchers can use this method to evaluate the impact of environmental factors like pollution and climate change on soil characteristics. For example, in the context of soil organic carbon pools in organic soils, PERMANOVA can be applied to assess the effects of different levels of pollution on carbon storage. By comparing the soil organic carbon pools across contaminated and uncontaminated sites, PERMANOVA helps to determine the extent to which pollution affects soil carbon sequestration, which is critical for climate change mitigation strategies (Yao et al. 2024).

Lastly, PERMANOVA's ability to handle high-dimensional data makes it ideal for evaluating soil properties that are influenced by multiple variables. In a study on soil organic carbon pools, researchers might measure various carbon fractions such as particulate organic carbon, mineral-associated organic carbon, and dissolved organic carbon. By using PERMANOVA, they can assess the overall differences in carbon pools between organic and conventional farming systems, considering all measured variables simultaneously. This holistic approach provides a deeper understanding of how different agricultural practices influence soil carbon dynamics, which is essential for developing sustainable land management practices (Yue and Hu 2022).

Software for PERMANOVA

Several software packages can perform PERMANOVA, including:

1. R (vegan package): Widely used in ecological research. The adonis function in the vegan package is commonly used for PERMANOVA.
2. PRIMER: Specialized software for ecological and environmental data analysis, which includes PERMANOVA.

3. PAST: Free software for statistical analysis of paleontological and ecological data, which includes PERMANOVA.

Example Application in Soil Science

Imagine you are studying the impact of different agricultural practices on soil microbial communities (Sun et al. 2024; Tang et al. 2016). You want to test if the microbial community composition differs between conventional and organic farming systems.

1. Research Question: Do soil microbial communities differ between conventional and organic farming systems?
2. Data Collection: Collect soil samples from both farming systems and perform DNA sequencing to identify microbial taxa.
3. Distance Measure: Use Bray-Curtis dissimilarity to quantify differences in microbial community composition between samples.
4. Create Distance Matrix: Compute the Bray-Curtis distance matrix from the sequencing data.
5. Perform PERMANOVA: Use software like R (with the vegan package) to perform the PERMANOVA test. Specify the distance matrix and the farming system as the grouping factor.

   Open any version of the R software

   Install vegan package

   Run the following scrips:

   ```
   library(vegan)
   dist_matrix<-vegdist(your_data, method = "bray")
   perm_result <- adonis(dist_matrix ~ farming_system, data = your_metadata)
   ```
6. Interpret Results: Check the pseudo-F statistic and p-value from the PERMANOVA output. A significant p-value indicates that microbial communities differ between the farming systems.

Conclusion

PERMANOVA (Permutational Multivariate Analysis of Variance) is a robust and versatile statistical tool that has become increasingly important in soil science for analyzing complex, multivariate data. Its non-parametric nature allows it to handle high-dimensional datasets without the strict assumptions required by traditional parametric tests, making it particularly suited for ecological and environmental studies where data often deviate from normality and homogeneity of variances. PERMANOVA's reliance on distance matrices and permutation techniques enables it to assess significant differences in multivariate datasets, such as soil microbial communities, physical properties, and organic carbon pools, across different land-use types and environmental conditions. This flexibility is crucial for soil scientists aiming to understand the intricate interactions and impacts of various factors on soil health and ecosystem functioning. By providing a comprehensive analysis of soil properties and microbial compositions, PERMANOVA facilitates the development of sustainable land management practices and informs strategies for mitigating environmental impacts, thus playing a critical role in advancing soil science research and contributing to the broader field of environmental science.

References

Anderson MJ (2017) Permutational Multivariate Analysis of Variance (PERMANOVA). In: Balakrishnan N, Colton T, Everitt B, Piegorsch W, Ruggeri F, Teugels JL (eds) Wiley StatsRef: Statistics Reference Online. Wiley. https://doi.org/10.1002/9781118445112.stat07841

Anderson MJ, Walsh DCI (2013) PERMANOVA, ANOSIM, and the Mantel test in the face of heterogeneous dispersions: what null hypothesis are you testing? Ecol Monogr 83:557–574. https://doi.org/10.1890/12-2010.1

Botha D, Barnard S, Claassens S, Rajakaruna N, Venter A, Ismail A, Allam M, Siebert SJ (2024) Soil type and precipitation level have a greater influence on fungal than bacterial diversity in serpentine and non-serpentine biological soil crusts. Ecol Res. https://doi.org/10.1111/1440-1703.12500

Canini F, Adams BJ, D'Acqui LP, D'Alò F, Zucconi L (2024) Antarctic rock and soil microbiomes: shared taxa, selective pressures, and extracellular DNA effects. Geoderma 446:116918. https://doi.org/10.1016/j.geoderma.2024.116918

Dietrich M, Panhölzl C, Angel R et al (2024) Plant roots affect free-living diazotroph communities in temperate grassland soils despite decades of fertilization. Commun Biol 7:846. https://doi.org/10.1038/s42003-024-06522-w

Forsmark B, Bizjak T, Nordin A, Rosenstock NP, Wallander H, Gundale MJ (2024) Shifts in microbial community composition and metabolism correspond with rapid soil carbon accumulation in response to 20 years of simulated nitrogen deposition. Sci Total Environ 918:170741. https://doi.org/10.1016/j.scitotenv.2024.170741

Jaeger ACH, Hartmann M, Conz RF, Six J, Solly EF (2024) Prolonged water limitation shifts the soil microbiome from copiotrophic to oligotrophic lifestyles in Scots pine mesocosms. Environ Microbiol Rep 16:13211. https://doi.org/10.1111/1758-2229.13211

Jiang Y, Wu Y, Di H, Tian X, Wang T, Wang R, Zhao J, Hu R, Shaaban M (2024) Impact of biocrust on soil nitrogen maintenance and microbial composition in citrus orchards under varying urea application rates. Appl Soil Ecol 201:105497. https://doi.org/10.1016/j.apsoil.2024.105497

Kelly BJ, Gross R, Bittinger K, Sherrill-Mix S, Lewis JD, Collman RG, Bushman FD, Li H (2015) Power and sample-size estimation for microbiome studies using pairwise distances and PERMANOVA. Bioinformatics 31:2461–2468. https://doi.org/10.1093/bioinformatics/btv183

Le Roux JJ, Leishman MR, Geraghty D et al (2024) The impacts of invasive African olive on native Australian legumes via altered soil conditions do not persist as legacy effects. Plant Soil 496:361–373. https://doi.org/10.1007/s11104-023-06366-z

Lejoly JDM, Quideau SA, Laganière J, Karst J, Martineau C, Samad A (2024) Earthworm cast microbiomes differ across soil types in northern forests. Appl Soil Ecol 200:105466. https://doi.org/10.1016/j.apsoil.2024.105466

Newman NS, Abbott CA, Brame JE, Cando-Dumancela C, Fickling NW, Liddicoat C, Robinson JM, Breed MF (2024) Childcare centre soil microbiomes are influenced by substrate type and surrounding vegetation condition. Sci Total Environ 927:172158. https://doi.org/10.1016/j.scitotenv.2024.172158

Peña-Gómez AI, Núñez-Rodríguez LX, Mora-Motta DA, Suárez JC, Silva-Olaya AM, Pereira APA, Ortiz-Morea FA (2024) Unveiling soil bacterial diversity in the Andes-Amazon transition zone: impacts of forest conversion to pasture. Appl Soil Ecol 201:105486. https://doi.org/10.1016/j.apsoil.2024.105486

Segura D, Jordaan K, Díez B et al (2024) Diversity and functionality of soil prokaryotic communities in antarctic volcanic soils: insights from penguin-influenced environments. Polar Biol. https://doi.org/10.1007/s00300-024-03273-6

Shulman HB, Aronson EL, Dierick D, Pinto-Tomás AA, Botthoff JK, Artavia-León A et al (2024) Leafcutter ants enhance microbial drought resilience in tropical forest soil. Environ Microbiol Rep 16:e13251. https://doi.org/10.1111/1758-2229.13251

Sun YQ, Zhu D, Wu Y et al (2024) Litter quality controls the earthworm microbiome in cropping fields. Plant Soil. https://doi.org/10.1007/s11104-024-06631-9

Tang Z-Z, Chen G, Alekseyenko AV (2016) PERMANOVA-S: association test for microbial community composition that accommodates confounders and multiple distances. Bioinformatics 32:2618–2625. https://doi.org/10.1093/bioinformatics/btw311

Wang X, Zhou L, Fu Y, Jiang Z, Jia S, Song B, Liu D, Zhou X (2024) Drought-induced changes in rare microbial community promoted contribution of microbial necromass C to SOC in a subtropical forest. Soil Biol Biochem 189:109252. https://doi.org/10.1016/j.soilbio.2023.109252

Yao T, Bucka F, Kögel-Knabner I, Knief C (2024) Organic matter type and soil texture shape prokaryotic communities during early-stage soil structure formation. J Plant Nutr Soil Sci 187:89–103. https://doi.org/10.1002/jpln.202300142

Yue Y, Hu Y-J (2022) Extension of PERMANOVA to testing the mediation effect of the microbiome. Genes 13:940. https://doi.org/10.3390/genes13060940

2 Analysis of Similarities (ANOSIM)

Abstract

ANOSIM (Analysis of Similarities) and ANOVA (Analysis of Variance) differ primarily in their approach and assumptions. While ANOVA compares means among groups assuming normally distributed data and homogeneity of variances, ANOSIM is a non-parametric test that evaluates differences in rank similarities between groups based on a distance matrix, making it suitable for ecological and environmental data that often do not meet ANOVA's assumptions. ANOSIM's main assumptions include the independence of samples and the appropriate choice of a distance measure reflecting ecological differences. It handles high-dimensional soil data effectively by operating on rank similarities within a distance matrix, allowing it to analyze multiple variables simultaneously without being constrained by their distribution. This capability makes ANOSIM particularly useful in soil science for studying diverse and complex datasets, such as microbial communities, soil properties, and environmental impacts. To ensure reliable results, researchers should check for independence of samples and choose a suitable distance measure that accurately represents the ecological relationships within the data before applying ANOSIM.

Keywords

Analysis of similarities · Environmental impacts · Microbial communities · Non-parametric test · R statistics

Questions Covered in the Chapter

(a) What is the primary difference between ANOSIM and traditional ANOVA?
(b) What are the main assumptions required for ANOSIM?
(c) How does ANOSIM handle high-dimensional ecological data?

T. Souza, *Advanced Statistical Analysis for Soil Scientists*,
https://doi.org/10.1007/978-3-031-88161-9_2

(d) Why is ANOSIM particularly useful in soil science research?
(e) What should researchers check for before applying ANOSIM to ensure reliable results?

Introduction

Analysis of Similarities (ANOSIM) is a non-parametric statistical test widely used in ecology and environmental sciences to compare the similarity between two or more groups of samples. ANOSIM operates on a matrix of dissimilarities, such as Bray-Curtis, Euclidean, or Jaccard distances, and assesses whether the observed dissimilarities between groups are significantly greater than those within groups. The test produces an R statistic, ranging from −1 to 1, where values close to 1 indicate complete separation between groups, values close to 0 suggest no difference between groups, and values below 0 indicate more similarity within groups than between them (Zhang et al. 2024).

The main assumptions of ANOSIM include the requirement that the data be independent and that the dissimilarities be appropriate for the data type and research question. ANOSIM does not assume normality, making it suitable for ecological data, which often do not follow normal distributions. Another assumption is that the sampling effort is comparable across groups to ensure valid comparisons. Additionally, the test assumes that the distance metric chosen accurately reflects the ecological differences among samples (Wang et al. 2024a).

One of the significant advantages of ANOSIM is its robustness to the lack of normality in data, making it an excellent choice for analyzing ecological data, which are often complex and non-normal. ANOSIM is also straightforward to interpret and implement, allowing researchers to draw meaningful conclusions about community differences with relative ease. Its flexibility in using different distance measures and its ability to handle multiple groups make it a versatile tool for advanced statistical analysis in ecology and environmental sciences (Segura et al. 2024).

In soil science, and particularly in soil microbiology, ANOSIM plays a crucial role in understanding the diversity and distribution of microorganisms, such as Basidiomycota, across different soil samples. For example, ANOSIM can be used to compare the fungal community composition in soils from different forest types or management practices in the Amazon rainforest. By identifying significant differences in microbial communities, researchers can infer the effects of environmental factors, such as soil pH, moisture, and nutrient availability, on the distribution of Basidiomycota. This understanding is essential for soil health assessment and the development of sustainable soil management practices (Peng et al. 2024).

Main Differences Between ANOSIM and ANOVA

ANOSIM (Analysis of Similarities) and traditional ANOVA (Analysis of Variance) are both statistical methods used to test hypotheses, but they differ significantly in their application and underlying assumptions. ANOVA is primarily used for comparing the means of different groups to determine if there are statistically significant differences among them. It assumes that the data are normally distributed and that variances are homogeneous across groups. ANOVA is often applied to continuous, normally distributed data and works well when these assumptions are met (Liu et al. 2024).

In contrast, ANOSIM is a non-parametric method used to test whether there is a significant difference between two or more groups based on a ranked distance matrix. Unlike ANOVA, ANOSIM does not require the data to follow a specific distribution, making it suitable for non-normally distributed data or data that do not meet the homogeneity of variance assumption. ANOSIM is often used in ecological studies where the focus is on comparing community composition rather than comparing means of individual variables (Peng et al. 2024).

In the field of soil science, particularly when studying soil bacteria, ANOVA might be used to compare the abundance of a specific bacterial species across different soil treatments. For example, if researchers are investigating the effect of different fertilizers on the population size of a particular soil bacterium, ANOVA can help determine if the mean population size differs significantly between treatment groups. This approach is useful when the interest lies in a single variable measured across groups (Boutsika et al. 2024).

On the other hand, ANOSIM would be more appropriate for analyzing overall community composition of soil bacteria under different environmental conditions or treatments. For instance, if researchers are interested in understanding how the entire bacterial community structure varies between soils treated with organic versus inorganic fertilizers, they would use ANOSIM to compare the similarity of bacterial communities across these treatments. ANOSIM would provide insights into whether the bacterial communities in the different treatments are significantly different from each other based on a chosen distance metric, such as Bray-Curtis dissimilarity. This non-parametric approach is advantageous in handling the complex and often non-normally distributed data typical of ecological community studies (Dong et al. 2024).

Key Concepts of ANOSIM

ANOSIM begins by constructing a distance matrix to quantify the dissimilarity between pairs of samples. In soil science, this could involve comparing chemical properties such as pH, nutrient content, or organic matter between soil samples from different plots. Common distance measures used include Bray-Curtis, which considers species abundance data, Euclidean for straightforward geometric distance, and Jaccard for presence-absence data. For example, in studying mycorrhizal associations, a distance matrix might be created based on the composition and abundance of mycorrhizal species in soils under different agricultural practices (Wang et al. 2024b).

Once the distance matrix is constructed, ANOSIM ranks all pairwise distances within and between groups. It compares the average rank of distances between groups to the average rank of distances within groups. For instance, when comparing the archaeal communities in soils from natural forests versus agricultural lands, ANOSIM would rank the dissimilarities in community composition. If the distances (dissimilarities) within the agricultural group and within the forest group are smaller than those between the two groups, this would suggest distinct archaeal communities in each environment (Xiang et al. 2024).

The core output of ANOSIM is the R statistic, which ranges from −1 to 1. An R value close to 1 indicates high dissimilarity between groups and high similarity within groups, suggesting strong group differentiation. An R value around 0 indicates no significant difference between groups, while an R value less than 0 (rare in practice) suggests more similarity between groups than within groups. For example, if comparing the dsDNA content, which indicates total microbial biomass, in soils from organic farms versus conventional farms, a high R value would suggest significant differences in microbial biomass between the farming practices (Wang et al. 2024c).

To assess the significance of the R statistic, ANOSIM uses permutations. This involves randomly shuffling the sample labels many times to create a distribution of R values under the null hypothesis of no group differences. For example, in studying the chemodiversity of organic molecules in soils from different land uses, permutations would help determine whether observed differences in chemodiversity between natural and agricultural soils are statistically significant or could have occurred by chance. This step provides a robust method for confirming the significance of the observed group differences.

By understanding these key concepts, young soil scientists can effectively apply ANOSIM to analyze complex ecological data, such as comparing soil chemical properties, mycorrhizal associations, archaeal communities, dsDNA content, and chemodiversity in various soil environments. This method provides a powerful tool for detecting and interpreting differences in soil communities and properties across different conditions (Zhao et al. 2024).

To Summarize

1. Distance Matrix: ANOSIM uses a distance matrix to quantify the dissimilarity between pairs of samples. Common distance measures include Bray-Curtis, Euclidean, and Jaccard distances.
2. Ranked Distances: ANOSIM ranks all pairwise distances within and between groups. It then compares the average rank of distances between groups to the average rank of distances within groups.
3. R Statistic: The ANOSIM test produces an R statistic, which ranges from −1 to 1:

   $R \approx 1$: High dissimilarity between groups, with more similarity within groups.

   $R \approx 0$: No significant difference between groups.

   $R < 0$: More similarity between groups than within groups (rare in practice).
4. Permutations: ANOSIM assesses the significance of the R statistic through permutations. The sample labels are randomly shuffled many times to generate a distribution of R under the null hypothesis of no group differences.

Main Assumptions Required for ANOSIM

The first assumption for ANOSIM is that the chosen dissimilarity metric appropriately reflects the differences between samples. This metric should capture the ecological or biological dissimilarity relevant to the study. In soil science, when comparing agroforestry systems, integrated crop-livestock systems, and conventional farming systems, the Bray-Curtis dissimilarity might be used to quantify differences in microbial community composition or soil organic matter. The suitability of the metric ensures that the observed differences are meaningful in the context of the study (Yu et al. 2024).

Secondly, ANOSIM assumes that the variance in dissimilarities is similar across groups. This means that the spread or variability of dissimilarity values within each group should be comparable. For example, when comparing soil properties such as nutrient content or microbial diversity across agroforestry, integrated crop-livestock, and conventional farming systems, the variability within each farming system should be similar. If one group has much higher variance, it could affect the interpretation of the ANOSIM results, potentially leading to misleading conclusions about the differences between groups (Qin et al. 2024).

Next, while ANOSIM is relatively robust to differences in group sizes, extremely unbalanced group sizes can impact the power and reliability of the test. Ideally, the number of samples in each group should be reasonably balanced. For instance, when studying soil microbial communities across different farming systems, having a similar number of soil samples from agroforestry plots, integrated crop-livestock plots, and conventional plots helps ensure that the test's conclusions are not unduly influenced by the sample size of any one group. This balance enhances the comparability of the results (Salas et al. 2024).

Finally, ANOSIM assumes that samples are independent of each other, meaning that the samples within and between groups should not be spatially or temporally autocorrelated. In practical terms, soil samples taken from agroforestry systems, integrated crop-livestock systems, and conventional farming systems should be collected in such a way that they represent independent observations. For example, if sampling soil from different plots within the same system, ensuring sufficient spatial separation between samples helps meet this assumption. Independence ensures that the observed dissimilarities are due to differences in the groups being compared rather than inherent similarities among closely located samples (Mondal et al. 2024).

Steps to Conduct ANOSIM

To conduct an ANOSIM, start by defining a clear research question. For instance, you might hypothesize that different land management practices, such as organic farming versus conventional farming, result in distinct soil microbial communities. This initial step ensures that your study has a focused objective, such as assessing whether the composition of Archaea, bacterial, and fungal communities differs significantly between organic and conventional soils (Lin et al. 2024).

Next, collect comprehensive multivariate data relevant to your hypothesis. In soil science, this could involve gathering soil samples from both organic and conventional fields and measuring various properties, such as pH, nutrient levels, and microbial community compositions. The microbial data might include sequencing results that provide information on the abundance and diversity of Archaea, bacteria, and fungi present in each sample. This step is crucial for obtaining the detailed information needed to compare the different soil communities (Islam et al. 2024a, b).

After collecting your data, choose an appropriate distance measure that fits your data type and research question. For microbial community data, Bray-Curtis dissimilarity is a common choice as it accounts for the abundance and presence-absence of species. Once you've selected your distance measure, compute the distance matrix from your multivariate data. This matrix will quantify the dissimilarities between all pairs of samples, reflecting the differences in microbial community composition between the soils under different management practices (Zhang et al. 2024).

Finally, use software to perform the ANOSIM test, specifying the distance matrix and the grouping factor, which in this case is the type of land management practice (organic versus conventional). Once the test is run, interpret the results by assessing the R statistic and the associated p-value. A high R value close to 1 would indicate significant differences between the microbial communities in organic and conventional soils, while a low R value near 0 would suggest no substantial differences. This process allows you to determine if the management practices have a significant impact on the soil microbial communities, providing valuable insights into soil health and sustainability (Wang et al. 2024b).

To Summarize

1. Define the Research Question: Clearly specify the hypothesis you want to test, such as differences in soil properties under different land management practices.
2. Collect Data: Gather multivariate data. In soil science, this might include measurements of various soil properties (e.g., pH, nutrient levels) or microbial community compositions.
3. Choose a Distance Measure: Select an appropriate distance measure based on your data type and research question.
4. Create a Distance Matrix: Compute the distance matrix from your multivariate data.
5. Perform ANOSIM: Use software to conduct the ANOSIM test. This involves specifying the distance matrix and the grouping factor (e.g., treatment type).
6. Interpret Results: Assess the R statistic and the associated p-value to determine if there are significant differences between the groups.

Handling High-Dimensional Ecological Data with ANOSIM

ANOSIM is well-suited for high-dimensional data, such as the complex datasets often encountered in soil science. This capability is particularly valuable when studying agroecological systems, where multiple soil properties and microbial community compositions need to be analyzed simultaneously. The method starts by converting high-dimensional data into a single distance matrix that captures the dissimilarities between all pairs of samples. This matrix simplifies the complexity, allowing for the comparison of groups based on the overall differences in their soil properties and microbial communities (Wang et al. 2024a, b, c).

For instance, in agroecological systems, researchers might measure a variety of soil properties, including pH, nutrient levels, organic matter content, and microbial communities (e.g., Archaea, bacteria, and fungi). These measurements create a high-dimensional dataset that can be difficult to analyze with traditional methods. By using ANOSIM, these diverse data types are transformed into a distance matrix using an appropriate dissimilarity metric, such as Bray-Curtis for microbial data or Euclidean distance for soil chemical properties. This matrix effectively summarizes the complex data, making it manageable for further analysis (Peng et al. 2024; Mondal et al. 2024).

Once the distance matrix is created, ANOSIM ranks the distances and compares the average ranks of distances within groups (e.g., different agroecological practices) to those between groups. This step helps to determine if the soil properties and microbial communities within each practice are more similar to each other than to those from different practices. For example, if comparing soils from agroforestry systems to those from monoculture farming, ANOSIM would assess whether the overall soil composition is more homogeneous within each system than between them.

The final step in ANOSIM involves interpreting the R statistic and the p-value to determine the significance of the observed differences. In the context of agroecological systems, a significant R statistic would indicate that the soil properties and microbial communities are distinctly different between the compared systems. This insight can help identify the impact of various farming practices on soil health and guide the development of sustainable agricultural practices that promote biodiversity and soil quality. ANOSIM's ability to handle high-dimensional data makes it a powerful tool for analyzing the complex interactions within agroecological systems, providing clear and actionable results.

Why ANOSIM Is Particularly Useful in Soil Science Research?

ANOSIM (Analysis of Similarities) is particularly useful in soil science research because it provides a robust method for comparing differences in community structure between groups, which is crucial for understanding the impacts of various factors on soil ecosystems. ANOSIM is non-parametric and operates on a distance matrix, making it suitable for ecological data that often violate the assumptions of parametric tests (Islam et al. 2024b). Here are some specific examples of its application in soil science, particularly in the context of biological invasion:

(a) Assessing Soil Microbial Community Changes: Biological invasions can significantly alter the composition and structure of soil microbial communities. ANOSIM can be used to compare microbial communities in invaded and non-invaded soils. For example, in a study examining the impact of an invasive plant species on soil bacteria, ANOSIM can help determine if there are significant differences in microbial community composition between soils with and without the invasive species. This provides insights into how invasions disrupt microbial ecosystems and soil health (Boutsika et al. 2024).
(b) Evaluating Soil Fauna Diversity: Invasive species can also affect soil fauna such as nematodes, insects, and earthworms. ANOSIM can compare the diversity and abundance of soil fauna between invaded and non-invaded areas. For instance, researchers might use ANOSIM to analyze the differences in earthworm communities in soils dominated by native vegetation versus those invaded by non-native plants. This helps in understanding the broader ecological impacts of biological invasions on soil fauna diversity and functioning (Islam et al. 2024a, b).
(c) Investigating Soil Chemical Properties: Biological invasions can lead to changes in soil chemical properties, such as nutrient levels, pH, and organic matter content. ANOSIM can be applied to investigate whether there are significant differences in soil chemical properties between invaded and non-invaded sites. For example, a study might use ANOSIM to compare the nutrient profiles of soils under native vegetation with those under invasive plant cover, revealing how invasions alter soil chemistry and nutrient cycling (Mondal et al. 2024).
(d) Studying Soil Physical Properties: Invasive species can modify soil physical properties like texture, bulk density, and water retention. ANOSIM can help analyze differences in these properties across invaded and non-invaded soils. For example, researchers might employ ANOSIM to assess changes in soil compaction and porosity caused by invasive plant root systems compared to native plants. This is crucial for understanding the physical impacts of invasions on soil structure and its implications for plant growth and soil health (Qin et al. 2024).

ANOSIM‘s strength lies in its ability to handle complex, multivariate ecological data without the stringent assumptions required by parametric tests. By using a distance matrix to quantify dissimilarities between samples, ANOSIM can effectively compare the overall community structure or property profiles of different soil groups. This flexibility makes it an invaluable tool in soil science research, particularly for studying the multifaceted impacts of biological invasions on soil ecosystems.

Practical Application in Agroecological vs. Conventional Farming Systems

In agroecological and conventional farming systems, ANOSIM provides a powerful tool for comparing the differences in soil characteristics and biological communities. This statistical method is especially useful for analyzing complex ecological data that do not meet the assumptions of parametric tests. By using distance matrices to quantify dissimilarities between samples, ANOSIM allows researchers to assess whether the observed differences between farming systems are statistically significant, thereby offering insights into the impacts of different agricultural practices on soil health and ecosystem functioning.

In a comparative study of soil microbial communities between agroecological and conventional farming systems, ANOSIM can reveal significant differences in microbial composition and diversity. For example, agroecological systems often promote greater microbial diversity due to practices like crop rotation, organic amendments, and reduced chemical inputs. By constructing a distance matrix using measures such as Bray-Curtis dissimilarity, researchers can use ANOSIM to test whether the microbial communities in agroecological soils are significantly different from those in conventional soils. This analysis can help demonstrate the benefits of agroecological practices in enhancing soil microbial health, which is crucial for nutrient cycling and soil fertility.

ANOSIM can also be applied to compare soil chemical properties between these two farming systems. Conventional farming often relies heavily on synthetic fertilizers and pesticides, which can lead to soil degradation and nutrient imbalances. In contrast, agroecological systems typically use organic fertilizers and sustainable soil management practices, leading to different chemical profiles. By analyzing the soil chemical properties such as pH, nutrient levels, and organic matter content, ANOSIM can help determine whether there are significant differences between the farming systems. This comparison is essential for understanding how agroecological practices contribute to maintaining or improving soil chemical health and sustainability (Xiang et al. 2024).

Furthermore, ANOSIM is valuable for assessing soil physical properties in agroecological versus conventional farming systems. Soil physical properties like texture, bulk density, and water retention are critical for plant growth and soil health. Conventional farming practices, such as heavy machinery use and monocropping, can negatively impact these properties, leading to soil compaction and erosion. Agroecological practices, on the other hand, often enhance soil structure and resilience. By using ANOSIM to compare the physical properties of soils under different farming practices, researchers can identify significant differences and advocate for sustainable soil management practices that preserve soil integrity and productivity (Salas et al. 2024).

The practical application of ANOSIM in comparing agroecological and conventional farming systems provides a comprehensive approach to evaluating the multifaceted impacts of agricultural practices on soil ecosystems. By addressing microbial, chemical, and physical properties, ANOSIM helps to elucidate the benefits of agroecological practices in promoting soil health and sustainability. This analysis is crucial for guiding policy and management decisions aimed at fostering sustainable agricultural systems that support both environmental health and agricultural productivity (Peng et al. 2024).

Applying ANOSIM in Soil Science

ANOSIM is a valuable tool in soil science for analyzing complex, multivariate data. This statistical method is particularly useful for studying the diversity and composition of soil microbial communities, which play a crucial role in soil health and nutrient cycling. For instance, researchers can use ANOSIM to compare microbial communities across different land-use types or management practices. In a study comparing organic and conventional farming systems, ANOSIM can reveal significant differences in bacterial community compositions. Organic farming practices, which often include the use of compost and reduced chemical inputs, tend to support a more diverse and stable microbial community compared to conventional practices that rely heavily on synthetic fertilizers and pesticides. This insight is essential for understanding how different agricultural practices affect soil microbial health and function (Xiang et al. 2024).

In addition to microbial communities, ANOSIM can be used to assess differences in soil chemical properties across various sites or treatments. Soil chemical properties, such as nutrient levels, pH, and organic matter content, are vital indicators of soil fertility and health. By applying ANOSIM, researchers can compare the chemical profiles of soils from different management practices or land uses. For example, a study might examine soils from agroforestry systems versus monoculture cropping systems. ANOSIM can help determine whether the soil nutrient levels and organic matter content differ significantly between these systems, highlighting the benefits of agroforestry in maintaining soil fertility and promoting sustainable agricultural practices (Yu et al. 2024).

ANOSIM is also useful for analyzing soil physical properties, which are critical for understanding soil structure, water retention, and erosion potential. Soil physical properties, such as texture, bulk density, and porosity, can vary significantly under different land management practices. Researchers can use ANOSIM to compare these properties across different sites or treatments to identify significant differences. For example, a study comparing soils under no-till farming with those under conventional tillage might find that no-till practices lead to better soil structure and reduced compaction. ANOSIM helps quantify these differences, providing valuable insights into how different farming practices impact soil physical health and resilience.

Environmental impact studies benefit greatly from ANOSIM as well, particularly when evaluating the effects of pollution, climate change, and other environmental factors on soil characteristics. For example, researchers can use ANOSIM to assess the impact of heavy metal contamination on soil organic carbon pools in organic soils. By comparing contaminated and uncontaminated sites, ANOSIM can reveal significant differences in the quantity and quality of soil organic carbon. This analysis helps to understand the extent of pollution's impact on soil health and guides remediation efforts. Overall, ANOSIM's ability to handle complex, multivariate data makes it an essential tool for soil scientists aiming to uncover significant patterns and relationships in soil ecosystems.

Software for ANOSIM

Several software packages can perform ANOSIM, including:

1. R (vegan package): Widely used in ecological research. The anosim function in the vegan package is commonly used for ANOSIM.
2. PRIMER: Specialized software for ecological and environmental data analysis, which includes ANOSIM.
3. PAST: Free software for statistical analysis of paleontological and ecological data, which includes ANOSIM.

Example Application in Soil Science

Imagine you are studying the impact of different land management practices on soil microbial communities. You want to test if the microbial community composition differs between forest, grassland, and agricultural soils.

1. Research Question: Do soil microbial communities differ between forest, grassland, and agricultural soils?
2. Data Collection: Collect soil samples from each land management type and perform DNA sequencing to identify microbial taxa.
3. Distance Measure: Use Bray-Curtis dissimilarity to quantify differences in microbial community composition between samples.
4. Create Distance Matrix: Compute the Bray-Curtis distance matrix from the sequencing data.
5. Perform ANOSIM: Use software like R (with the vegan package) to perform the ANOSIM test. Specify the distance matrix and the land management type as the grouping factor.

   ```
   Open any version of the R software
   library(vegan)
   dist_matrix <- vegdist(your_data, method = "bray")
   anosim_result <- anosim(dist_matrix, your_metadata$land_management)
   ```
6. Interpret Results: Check the R statistic and p-value from the ANOSIM output. A significant p-value indicates that microbial communities differ between the land management types.

Conclusion

ANOSIM (Analysis of Similarities) is a vital tool for soil science and data analysis, particularly when dealing with complex and high-dimensional datasets. By effectively quantifying and comparing dissimilarities between soil samples, ANOSIM enables researchers to uncover significant differences in soil properties and microbial communities across various management practices and environments. For young scientists, mastering ANOSIM offers a robust method for interpreting intricate ecological and chemical data, enhancing their ability to address research questions related to soil health, sustainability, and ecosystem functioning. Its non-parametric nature and capacity to handle diverse types of data make ANOSIM an invaluable asset in advancing our understanding of soil systems and guiding the development of informed agricultural practices. As such, proficiency in ANOSIM not only strengthens analytical skills but also contributes to the generation of meaningful insights that can drive innovative solutions in soil science.

References

Boutsika A, Xanthopoulou A, Tanou G et al (2024) A microbiome survey of contrasting potato terroirs using 16S rRNA long-read sequencing. Plant Soil. https://doi.org/10.1007/s11104-024-06686-8

Dong S, Zhang B, Wang Z, Zhou X, Gao Q (2024) Responses of soil bacterial communities and maize yields to sulfur application across four soil types. Front Microbiol. https://doi.org/10.3389/fmicb.2024.1329938

Islam W, Zeng F, Almoallim HS, Ansari MJ (2024a) Unveiling soil animal community dynamics beneath dominant shrub species in natural desert environment: implications for ecosystem management and conservation. J Environ Manag 366:121697. https://doi.org/10.1016/j.jenvman.2024.121697

Islam W, Zeng F, Dar AA, Yousaf MS (2024b) Dynamics of soil biota and nutrients at varied depths in a Tamarix ramosissima-dominated natural desert ecosystem: implications for nutrient cycling and desertification management. J Environ Manag 354:120217. https://doi.org/10.1016/j.jenvman.2024.120217

Lin X, Han X, Yang J, Liu F, Li Y, Chen Z (2024) Network of soil fungi and the microfauna community under diverse anthropic disturbances under *Chrysopogon zizanioides* planting in the reservoir. Plants 13:393. https://doi.org/10.3390/plants13030393

Liu Z, Nan Z, Lin S, Meng W, Xie L, Yu H, Zhang Z, Wan S (2024) Peanut-based intercropping systems altered soil bacterial communities, potential functions, and crop yield. PeerJ 12:e16907. https://doi.org/10.7717/peerj.16907

Mondal S, Burgos-Hernandez T, Ralston TI et al (2024) Urban greenspaces shape soil nematode community across soil depth gradients: belowground life at The Ohio State University. Ecol Indic 166:112399. https://doi.org/10.1016/j.ecolind.2024.112399

Peng Q, Zheng H, Xu H et al (2024) Response of soil fungi to textile dye contamination. Environ Pollut 359:124577. https://doi.org/10.1016/j.envpol.2024.124577

Qin P, Hu L, Liu Y et al (2024) Responses of soil microbial communities and nutrient dynamics under continuous alfalfa (*Medicago sativa* L.) cultivation. Appl Soil Ecol 197:105356. https://doi.org/10.1016/j.apsoil.2024.105356

Salas A, Fusaro B, Rusconi JM et al (2024) Diversity and abundance of free-living nematodes from Carlini Station, 25 de Mayo/King George Island, Antarctica: a case study in pristine and disturbed soils. Polar Biol 47:73–83. https://doi.org/10.1007/s00300-023-03211-y

Segura D, Jordaan K, Díez B et al (2024) Diversity and functionality of soil prokaryotic communities in antarctic volcanic soils: insights from penguin-influenced environments. Polar Biol. https://doi.org/10.1007/s00300-024-03273-6

Wang W, Yang Y, Li J et al (2024a) Consecutive fertilization-promoted soil nutrient availability and altered rhizosphere bacterial and bulk fungal community composition. Forests 15:514. https://doi.org/10.3390/f15030514

Wang X, Zhang N, Chen K et al (2024b) Response mechanism of soil microorganisms to simulated precipitation in the source wetland of Qinghai Lake. Ecol Process 13:25. https://doi.org/10.1186/s13717-024-00502-y

Wang Y, Zhu K, Chen X, Wei K, Wu R, Wang G (2024c) Stochastic and deterministic assembly processes of bacterial communities in different soil aggregates. Appl Soil Ecol 193:105153. https://doi.org/10.1016/j.apsoil.2023.105153

Xiang J, Gu J, Wang G, Bol R, Yao L, Fang Y, Zhang H (2024) Soil pH controls the structure and diversity of bacterial communities along elevational gradients on Huangshan, China. Eur J Soil Biol 120:103586. https://doi.org/10.1016/j.ejsobi.2023.103586

Yu T, Hou X, Fang X, Razavi B, Zang H, Zeng Z, Yang Y (2024) Short-term continuous monocropping reduces peanut yield mainly via altering soil enzyme activity and fungal community. Environ Res 245:117977. https://doi.org/10.1016/j.envres.2023.117977

Zhang Z, Wang Z, Zhang T, Yin B, Li R, Sheng Z, Li S (2024) Variations in soil microbial communities in different saline soils under typical Populus spp. vegetation in alpine region of the Qaidam Basin, NW China. Ecotoxicol Environ Saf 282:116747. https://doi.org/10.1016/j.ecoenv.2024.116747

Zhao L, Wei D, Jiang C et al (2024) *Pseudostellaria heterophylla–Capsicum annuum* rotation improves soil acidification and bacterial communities in continuously cropped *Pseudostellaria heterophylla* fields. J Soil Sci Plant Nutr. https://doi.org/10.1007/s42729-024-01887-7

Multiple Regression

3

Abstract

Multiple regression extends traditional linear regression by allowing the analysis of multiple independent variables simultaneously, rather than just one. While traditional linear regression examines the relationship between a single predictor and an outcome, multiple regression evaluates how several predictors collectively influence the dependent variable. The main assumptions for multiple regression include linearity (the relationship between predictors and the outcome should be linear), independence (errors should be independent of each other), normality (errors should be normally distributed), and homoscedasticity (constant variance of errors). Additionally, addressing multicollinearity is crucial when dealing with high-dimensional soil data to prevent predictors from being excessively correlated. Multiple regression is particularly useful in soil science research for understanding complex interactions between various soil properties and their impact on outcomes like soil fertility and plant growth. Before applying multiple regression, researchers should ensure data quality by checking for missing values, outliers, and ensuring the data meets the necessary assumptions to obtain reliable and meaningful results.

Keywords

Assumptions · Independent variables · Multiple regression · Predictions · Soil science

Questions Covered in the Chapter

(a) What is the primary difference between multiple regression and traditional linear regression?
(b) What are the main assumptions required for multiple regression?

T. Souza, *Advanced Statistical Analysis for Soil Scientists*,
https://doi.org/10.1007/978-3-031-88161-9_3

(c) How does multiple regression handle high-dimensional soil data?
(d) Why is multiple regression particularly useful in soil science research?
(e) What should researchers check for before applying multiple regression to ensure reliable results?

Introduction

Multiple regression is a statistical technique that models the relationship between a dependent variable and two or more independent variables. This method extends simple linear regression, which involves only one predictor, by allowing for the simultaneous consideration of multiple factors (Shi et al. 2024). The general form of a multiple regression model is $Y = \beta_0 + \beta_1 X_1 + \beta_2 X_2 + \ldots + \beta_n X_n + \epsilon$, where Y is the dependent variable, β_0 is the intercept, β_1, β_2, …, β_n are the coefficients of the independent variables X_1, X_2, …, X_n and ϵ is the error term (Lipatov et al. 2024). This method is powerful for exploring and quantifying the impact of multiple predictors on a single outcome.

For multiple regression to provide valid results, several key assumptions must be met. Firstly, there should be a linear relationship between the dependent and independent variables (Khasawneh et al. 2024). Secondly, the residuals (errors) of the model should be normally distributed and independent of each other, which ensures that the model's estimates are unbiased and efficient (Polo-Mendoza et al. 2024). Thirdly, there should be homoscedasticity, meaning the variance of residuals should be constant across all levels of the independent variables (Zema et al. 2024). Lastly, multicollinearity, which occurs when independent variables are highly correlated with each other, should be minimal as it can distort the results (Muhammad et al. 2024). These assumptions are crucial for the integrity and reliability of the regression analysis.

One of the primary advantages of multiple regression as an advanced statistical analysis tool is its ability to handle complex data sets with multiple predictors (Al Khazaleh et al. 2024). This makes it invaluable for understanding and interpreting the multifaceted relationships within data, enabling researchers to isolate the effect of each independent variable while controlling for the influence of others (Wu et al. 2024). Additionally, multiple regression can improve the accuracy of predictions and the robustness of findings by incorporating a broader spectrum of information (Sanjeewani et al. 2024). Its versatility allows for the analysis of various types of data, including continuous, categorical, and binary variables, making it a comprehensive method for statistical analysis (Mozaffari et al. 2024).

In soil science, multiple regression is particularly important as it allows for the analysis of numerous factors that influence soil properties and processes (Kumar et al. 2024a). For example, soil fertility, moisture content, pH levels, and organic matter content can all be studied simultaneously to understand their collective impact on crop yields (Park et al. 2024). Multiple regression models can help in identifying key drivers of soil health and productivity, aiding in the development of effective soil management practices (Zahedifard et al. 2024). By providing insights into the complex interactions between various soil parameters, multiple regression supports the advancement of sustainable agriculture and environmental conservation efforts (Alzabeebee and Keawsawasvong 2024).

Main Differences Between Multiple Regression and Linear Regression

The primary difference between multiple regression and traditional linear regression lies in the number of independent variables each method uses to predict the dependent variable. Traditional linear regression, also known as simple linear regression, involves a single independent variable and its relationship with a single dependent variable. The model is represented as $Y = \beta_0 + \beta_1 X + \epsilon$, where Y is the dependent variable, X is the independent variable, β_0 is the intercept, β_1 is the slope, and ϵ is the error term. This method is suitable when the focus is on understanding the impact of one predictor on the outcome (Kumar et al. 2024b).

In contrast, multiple regression involves two or more independent variables to predict the dependent variable. The model is represented as $Y = \beta_0 + \beta_1 X_1 + \beta_2 X_2 + \ldots + \beta_n X_n + \epsilon$, where Y is the dependent variable, X_1, X_2, …, and X_n are the independent variables, and β_0, β_1, β_2, …, β_n are the coefficients associated with each predictor. Multiple regression is used when the outcome is influenced by several factors simultaneously, and it helps to understand the relative contribution of each predictor while controlling for the others (Kumar et al. 2024a).

For example, in the field of plant nutrition, a researcher might be interested in understanding how nitrogen (N) and phosphorus (P) levels affect plant growth. Using simple linear regression, the researcher could analyze the relationship between plant growth (the dependent variable) and nitrogen levels (the independent variable), resulting in a model like Plant Growth = $\beta_0 + \beta_1$(Nitrogen) + ϵ. This model would provide insights into how variations in nitrogen levels alone impact plant growth without considering the N and P interaction (Khasawneh et al. 2024).

However, plant growth is often influenced by multiple nutrients simultaneously. By using multiple regression, the researcher can incorporate both nitrogen and phosphorus levels as independent variables in the model. The resulting model would be Plant Growth = $\beta_0 + \beta_1$(Nitrogen) + β_2(Phosphorus) + ϵ. This approach allows the researcher to assess the combined effect of nitrogen and phosphorus on plant growth and understand how each nutrient independently contributes to the outcome while accounting for the presence of the other. This comprehensive analysis provides a more accurate and holistic understanding of the factors influencing plant growth (Al Khazaleh et al. 2024).

Key Concepts of Multiple Regression

When introducing young scientists to the key concepts of multiple regression, it's important to emphasize how this statistical technique can unravel complex relationships between variables. Multiple regression is a powerful tool that allows us to examine how several factors simultaneously influence an outcome. In soil science, this means understanding how different soil properties and nutrient levels interact to affect plant growth. For example, instead of looking at the effect of just one soil property on crop yields, multiple regression enables us to assess how a combination of properties—such as pH, organic matter, and nutrient levels—affects plant nutrition and growth (Egbueri and Khan 2024).

The fundamental idea behind multiple regression is to model the relationship between a dependent variable and multiple independent variables. This is done through an equation like $Y = \beta_0 + \beta_1X_1 + \beta_2X_2 + \ldots + \beta_nX_n + \epsilon$, where Y represents the outcome (e.g., plant growth), and $X_1, X_2, \ldots, X_n$ are the predictors (e.g., different soil chemical properties). By including multiple predictors, scientists can isolate the impact of each one while controlling for the effects of the others. This allows for a more nuanced understanding of how various factors contribute to the overall outcome (Al Khazaleh et al. 2024).

In practice, multiple regression can help scientists address questions such as how soil pH and nutrient levels together influence plant growth. For example, a researcher might use multiple regression to analyze how both nitrogen (N) and phosphorus (P) levels in the soil affect the yield of a specific crop. By including both nutrients in the model, they can determine not just the effect of each nutrient individually but also how their interaction influences plant health and productivity. This approach can reveal whether optimizing both nutrients simultaneously provides a greater benefit than adjusting each one independently.

Another application of multiple regression in soil science is assessing how different soil properties, such as texture, organic matter, and moisture, interact to affect soil fertility and plant growth. For instance, a scientist could model how soil texture and organic matter content together influence the availability of nutrients like potassium and calcium. This would help in understanding whether soils with high organic matter content mitigate the negative effects of poor texture on nutrient availability, guiding better soil management practices and fertilization strategies. By leveraging multiple regression, young scientists can gain deeper insights into the complex interactions that drive soil health and plant nutrition.

To Summarize

1. Definition and Purpose: The primary purpose is to understand how multiple factors contribute to or predict an outcome. For example, in soil science, you might use multiple regression to investigate how different soil properties collectively influence plant growth.

(continued)

2. Model Representation: The general form of a multiple regression model is $Y = \beta_0 + \beta_1X_1 + \beta_2X_2 + \ldots + \beta_nX_n + \epsilon$, where Y represents the outcome (e.g., plant growth), and X_1, X_2, …, X_n are the predictors (e.g., different soil chemical properties), and ϵ is the error term (representing the variation in Y not explained by the predictors).
3. Assumptions: For multiple regression results to be valid, certain assumptions must be met, like linearity, independence, normality, and homoscedasticity. Additionally, multicollinearity (high correlation between independent variables) should be minimal to avoid skewed results.
4. Applications and Benefits: Multiple regression is especially useful in scenarios where an outcome is influenced by several factors. For example, in soil science, it can help determine how combined levels of nitrogen, phosphorus, and soil moisture affect crop yield. By using multiple regression, scientists can make more accurate predictions, identify key drivers of plant growth, and develop more effective management strategies.

Main Assumptions Required for Multiple Regression

The main assumptions required for multiple regression are crucial for ensuring the accuracy and reliability of the model's results. These assumptions include linearity, independence, normality, and homoscedasticity. Each assumption plays a key role in validating the conclusions drawn from the regression analysis and in ensuring that the statistical tests applied are appropriate for the data (Alzabeebee and Keawsawasvong 2024).

Linearity is the first assumption and it stipulates that there should be a linear relationship between the dependent variable and each of the independent variables. This means that the effect of each predictor on the outcome should be additive and linear. In the context of soil science, if you are studying the impact of soil salinity and micronutrient levels on plant growth, the relationship between plant growth (the dependent variable) and each predictor (soil salinity and micronutrient levels) should be linear. For example, a multiple regression model might assess how increases in soil salinity and variations in zinc and copper levels affect crop yields. The assumption is that each increase in these predictors has a proportional effect on the yield (Mozaffari et al. 2024).

Independence of residuals is the second assumption, which requires that the residuals (the differences between observed and predicted values) be independent of each other. This implies that the value of one residual should not be influenced by the value of another. In soil science studies, this assumption is important when measurements are taken at different locations or times. For instance, if you're examining how soil salinity and micronutrient levels affect plant growth in various fields, the residuals from one field's observations should not be correlated with residuals from another. If residuals are correlated, it might indicate spatial or temporal autocorrelation, which could bias the results (Park et al. 2024).

Normality of residuals is the third assumption, which states that the residuals should be approximately normally distributed. This assumption is important for valid hypothesis testing and confidence intervals. For example, if you are modeling the effects of soil salinity and micronutrient levels on plant growth, the residuals from your regression analysis should follow a normal distribution. If residuals are not normally distributed, it could suggest that the model is not adequately capturing the relationship or that there are outliers affecting the results.

Homoscedasticity is the final assumption and refers to the requirement that the variance of residuals should be constant across all levels of the independent variables. This means that the spread of residuals should be roughly the same regardless of the values of the predictors. In practical terms, when studying how soil salinity and micronutrient levels impact plant growth, you should check that the residuals do not exhibit a pattern or change in spread as the values of salinity or micronutrients change. If residuals become more spread out or less consistent at different levels of predictors, this could indicate heteroscedasticity, which may require transforming variables or adjusting the model to correct for this issue (Mozaffari et al. 2024).

Steps to Conduct Multiple Regression

To conduct multiple regression analysis effectively, young scientists in soil science and plant nutrition should follow a systematic approach. The first step involves formulating a clear research question and hypothesis. For example, suppose you want to understand how soil physical properties such as texture and moisture, along with nutrients like nitrogen (N) and phosphorus (P), impact plant growth in organic soils. Begin by defining the dependent variable (e.g., plant growth) and independent variables (e.g., soil texture, moisture, N, and P levels). This step is crucial for setting up the analysis and ensuring that the model will address the specific scientific questions you are investigating.

The second step is to collect and prepare the data. Gather data on the dependent variable and all independent variables from soil and plant samples. In this case, you would need measurements of plant growth, soil texture (e.g., sand, silt, clay percentages), soil moisture content, and nutrient levels of N and P. Data preparation also involves checking for any missing values, outliers, or errors, and ensuring that the data meets the assumptions required for multiple regression analysis. Proper data cleaning and preparation are essential for obtaining reliable results (Al Khazaleh et al. 2024).

The third step is to perform the multiple regression analysis using statistical software. Input the cleaned data into the software and specify the dependent and independent variables. For instance, using a software tool, you would enter plant growth as the dependent variable and soil texture, moisture, N, and P levels as independent variables. The software will then calculate the regression coefficients and provide outputs such as R-squared values, coefficients, and p-values. These results will help you understand the strength and significance of the relationships between the predictors and the outcome.

Finally, interpret and validate the results of the regression analysis. Examine the coefficients to understand how each independent variable affects plant growth. For example, determine if soil moisture or nutrient levels have a significant impact on plant growth in organic soils. Assess the goodness-of-fit (e.g., R-squared value) to evaluate how well the model explains the variability in the dependent variable. Additionally, check for any violations of regression assumptions, such as linearity or multicollinearity, and adjust the model if necessary. This step ensures that your conclusions are valid and that the model provides meaningful insights into the relationships between soil properties, nutrients, and plant growth (Zema et al. 2024).

To Summarize

1. Formulate Research Question and Hypothesis:
   Define the dependent variable (e.g., plant growth).
   Identify independent variables (e.g., soil texture, moisture, N, and P levels).
   Develop a clear hypothesis on how these variables might interact.
2. Data Collection and Preparation:
   Gather relevant data on all variables (plant growth, soil properties, nutrient levels).
   Clean the data: address missing values, outliers, and errors.
   Ensure the data meets the assumptions required for multiple regression.
3. Perform Multiple Regression Analysis:
   Input the data into statistical software.
   Specify dependent and independent variables.
   Run the regression analysis to obtain coefficients, R-squared values, and p-values.
4. Interpret and Validate Results:
   Analyze the regression coefficients to understand variable impacts.
   Evaluate the model's goodness-of-fit (e.g., R-squared value).
   Check for violations of assumptions and adjust the model as needed.

Handling High-Dimensional Soil Data with Multiple Regression

Multiple regression is well-suited for handling high-dimensional soil data by allowing the inclusion of numerous predictors in a single analysis. For instance, if you're investigating how essential macronutrients like nitrogen (N), phosphorus (P), potassium (K), calcium (Ca), and magnesium (Mg) affect crop yield, multiple regression enables you to incorporate all these nutrients simultaneously into the model. This comprehensive approach helps in understanding the relative contributions of each nutrient while controlling for the effects of the others. The regression model might be expressed as: $\text{Crop Yield} = \beta_0 + \beta_1(N) + \beta_2(P) + \beta_3(K) + \beta_4(Ca) + \beta_5(Mg) + \epsilon$.

One challenge with high-dimensional data is multicollinearity, where independent variables are highly correlated, making it difficult to isolate the effect of each predictor. To address this, techniques such as Variance Inflation Factor (VIF) analysis can identify and mitigate multicollinearity. Additionally, regularization methods like Ridge or Lasso regression can help manage the complexity of the model and improve its interpretability by penalizing excessive coefficients or selecting only the most relevant predictors (Shi et al. 2024).

When dealing with numerous predictors, variable selection methods and regularization are essential for refining the model. For example, stepwise regression or regularization techniques can help determine which macronutrients most significantly impact crop yield and reduce the model's complexity. This can involve identifying key nutrients and potentially eliminating less influential ones, which enhances the model's performance and interpretability.

Ultimately, multiple regression allows for a nuanced analysis of how combinations of macronutrients interact to affect soil fertility and crop yield. By applying this method, researchers can gain insights into the most effective nutrient levels and combinations for optimizing agricultural productivity. For example, the model might show that while high levels of N and P are critical, the presence of adequate K, Ca, and Mg further enhances the effectiveness of N and P, leading to more targeted and efficient soil management practices (Wu et al. 2024).

Why Multiple Regression Is Particularly Useful in Soil Science Research?

Multiple regression is particularly useful in soil science research because it allows for the analysis of complex interactions between various soil properties and their combined effects on outcomes such as soil fertility, plant growth, and environmental health. This method provides a robust framework for understanding how multiple factors simultaneously influence a single outcome, which is crucial in the context of soil science where multiple variables are often at play (Polo-Mendoza et al. 2024).

For example, in the study of saline soils, multiple regression can be used to understand how different soil properties interact to affect plant growth. Saline soils, characterized by high concentrations of soluble salts, can be influenced by various factors such as soil texture, organic matter content, and the levels of specific ions like sodium (Na) and chloride (Cl). By using multiple regression, researchers can create a model to simultaneously evaluate how these factors contribute to plant stress and yield. A typical model might look like this: Plant Yield = $\beta_0 + \beta_1$(Soil Texture) + β_2(Organic Matter) + β_3(Na) + β_4(Cl) + ϵ.

In another example, multiple regression can help in managing salinity in agricultural practices. Researchers might use multiple regression to analyze how different soil amendments (such as gypsum or organic compost) and irrigation practices interact with the levels of soluble salts in the soil to affect crop productivity. The model could include predictors such as the type and amount of soil amendment, irrigation frequency, and initial salinity levels, helping to identify the most effective strategies for mitigating soil salinity and improving crop yields (Sanjeewani et al. 2024).

Additionally, multiple regression can be applied to assess the impact of soil salinity on various soil health indicators. For instance, a study might examine how soil salinity, texture, and moisture content collectively influence soil microbial activity or nutrient availability. By including these multiple factors in a regression model, researchers can gain insights into how soil salinity interacts with other soil properties to affect soil health and fertility, leading to better soil management practices.

Practical Application in Conventional Farming Systems Using Fertilizers' Levels

Multiple regression is a powerful tool for optimizing fertilizer use in conventional farming systems. In these systems, the goal is often to maximize crop yields while minimizing input costs and environmental impact. By applying multiple regression, farmers can analyze how different factors, such as fertilizer doses and soil properties, interact to influence crop production. For example, a researcher might use multiple regression to study how varying doses of nitrogen (N), phosphorus (P), and potassium (K) fertilizers affect crop yield. The model would include these fertilizer variables as predictors, allowing for the assessment of their combined impact on yield while accounting for other influencing factors such as soil texture and moisture (Al Khazaleh et al. 2024).

In practice, this means that multiple regression can help determine the optimal fertilizer combinations and application rates for different soil types. Suppose a farm has several fields with varying soil textures and moisture levels. By using multiple regression, a farmer can create a model that predicts crop yield based on different fertilizer doses and soil characteristics. This approach allows for the identification of the most effective fertilizer rates for each field, helping to avoid over-fertilization and under-fertilization. For instance, the model might reveal that fields with sandy soils require higher phosphorus doses compared to loamy soils to achieve optimal yields.

Moreover, multiple regression can be used to fine-tune fertilizer recommendations based on historical data. By analyzing past crop performance data alongside fertilizer application records, researchers can build a regression model to understand how different doses and combinations of fertilizers have historically impacted yields. This historical perspective helps in making data-driven decisions for future fertilization strategies. For example, if the model indicates that increasing potassium levels leads to significantly higher yields in certain conditions, farmers can adjust their fertilizer practices accordingly to enhance productivity.

Overall, the practical application of multiple regression in conventional farming systems provides valuable insights into how various fertilizer doses interact with soil characteristics to affect crop yields. By leveraging this analysis, farmers can optimize their fertilization practices, improving crop performance while managing input costs and reducing environmental impact. This data-driven approach ensures that fertilizer applications are tailored to specific field conditions, ultimately leading to more efficient and sustainable farming practices.

Applying Multiple Regression in Soil Science

Applying multiple regression in soil science offers a powerful method for understanding and managing the complex interactions between soil properties and plant nutrition. For young scientists in the field, it's essential to grasp how this technique can be used to analyze various soil factors and their combined effects on soil health and plant growth. Multiple regression allows researchers to assess how multiple soil characteristics—such as physical properties, chemical properties, and organic carbon pools—interact to influence outcomes like soil fertility and crop productivity (Egbueri and Khan 2024).

In practice, multiple regression can be used to evaluate how soil physical properties, such as texture and bulk density, impact soil health and plant growth. For example, if you want to understand how soil texture (proportions of sand, silt, and clay) and bulk density affect the availability of nutrients and plant growth in sandy soils, you can use multiple regression to model these relationships. The regression model might include predictors like soil texture and bulk density, with plant growth or nutrient availability as the dependent variable. This helps in identifying which physical properties most significantly affect soil fertility and optimizing soil management practices (Lipatov et al. 2024).

Additionally, multiple regression can analyze the influence of soil chemical properties on plant nutrition. Consider a scenario where you want to study how soil pH, nutrient levels (such as nitrogen, phosphorus, and potassium), and soil salinity interact to affect plant growth. By including these variables in a multiple regression model, you can assess their combined impact on plant health. For example, the model might reveal that while soil pH and phosphorus levels are critical for plant growth, their effects are moderated by soil salinity levels, providing insights into how to manage these chemical properties for optimal crop performance (Mozaffari et al. 2024).

Furthermore, multiple regression is valuable for studying soil organic carbon (SOC) pools, which are crucial for soil fertility and carbon sequestration. In sandy soils, which often have lower organic carbon content, multiple regression can be used to investigate how factors like land use, organic matter inputs, and soil management practices influence SOC levels. By building a regression model with these predictors, researchers can determine the key factors driving changes in SOC pools and develop strategies to enhance soil carbon storage. For instance, the model might show that increasing organic matter inputs significantly boosts SOC levels, helping to inform practices that improve soil health and sustainability.

Software for Multiple Regression

There are several software programs available for performing multiple regression analysis, each with its own strengths and features. Here's an overview of some popular options:

1. **R**: R is a free and open-source programming language specifically designed for statistical computing and graphics. It is widely used for multiple regression analysis due to its flexibility and extensive package ecosystem. In R, you can perform multiple regression using built-in functions like lm() (linear model) for fitting the regression model. R's extensive libraries, such as caret, glmnet, and MASS, provide additional tools for data preprocessing, model selection, and regularization. R is particularly favored for its powerful statistical analysis capabilities and customizable features.
2. **Python**: Python is another versatile and widely-used programming language that supports multiple regression analysis through various libraries. The statsmodels library provides functions for fitting and analyzing multiple regression models, including tools for assessing model assumptions and diagnostics. The scikit-learn library offers additional functionalities for machine learning and model evaluation. Python's integration with data manipulation libraries like pandas and visualization libraries like matplotlib makes it a comprehensive tool for data analysis.
3. **SPSS (Statistical Package for the Social Sciences)**: SPSS is a commercial software package used for statistical analysis in social science research. It features a user-friendly graphical interface that simplifies the process of performing multiple regression analysis. SPSS provides options for regression diagnostics, model building, and hypothesis testing. It is well-suited for users who prefer a point-and-click interface over programming.
4. **SAS (Statistical Analysis System)**: SAS is a powerful commercial software suite used for advanced analytics, including multiple regression analysis. It offers extensive features for data manipulation, statistical modeling, and reporting. The PROC REG procedure in SAS is used for performing linear regression, including multiple regression analysis. SAS is known for its robustness and is commonly used in large-scale data analysis and industry applications.
5. **Excel**: Microsoft Excel is a widely used spreadsheet software that also offers basic statistical analysis capabilities. The Data Analysis Toolpak add-in provides options for performing multiple regression analysis, including regression diagnostics and output analysis. While Excel is accessible and easy to use for basic regression tasks, it lacks some of the advanced features and flexibility offered by specialized statistical software.
6. **Minitab**: Minitab is a statistical software package designed for data analysis and quality improvement. It provides a user-friendly interface for performing multiple regression analysis and includes features for regression diagnostics, model fitting, and interpretation. Minitab is often used in educational settings and by professionals for its ease of use and comprehensive support for statistical methods.

Each of these software options has its own strengths and is suited for different types of users and applications. Choosing the right tool depends on your specific needs, familiarity with programming or software interfaces, and the complexity of your regression analysis.

Example Application in Soil Science

Multiple regression is a valuable tool in soil science, particularly for understanding how various soil properties interact to influence outcomes such as plant growth, soil health, and nutrient availability. One notable application is in assessing how soil physical properties affect crop yield. For instance, a researcher might want to determine how soil texture (proportions of sand, silt, and clay) and bulk density influence the productivity of crops in different soil types. By using multiple regression, the researcher can model crop yield as a function of these soil properties, enabling the identification of which factors are most significant and how they interact to impact crop performance. This approach helps in tailoring soil management practices to improve yields (Mozaffari et al. 2024).

In another application, multiple regression can be used to analyze the relationship between soil chemical properties and nutrient availability. For example, suppose you are studying how soil pH, nitrogen (N), phosphorus (P), and potassium (K) levels interact to affect plant nutrient uptake. By including these variables in a multiple regression model, you can assess how each chemical property contributes to nutrient availability and plant health. This model might reveal, for instance, that while soil pH is a crucial determinant of nutrient availability, the combined levels of N, P, and K significantly influence plant growth, providing insights into optimal fertilization strategies (Kumar et al. 2024a, b).

Additionally, multiple regression can help in understanding the dynamics of soil organic carbon (SOC) pools and their impact on soil fertility. In sandy soils, which often have lower organic carbon content, multiple regression can be used to examine how factors such as land use, organic matter inputs, and soil management practices affect SOC levels. By creating a regression model with these predictors, researchers can determine which factors are most influential in increasing SOC and improving soil fertility. For example, the model might show that organic matter inputs and specific land management practices significantly boost SOC levels, informing practices that enhance soil health.

Multiple regression is also useful for studying the effects of soil salinity on plant growth. In saline soils, high concentrations of soluble salts can impact plant health and productivity. By incorporating variables such as soil salinity, soil texture, and organic matter content into a multiple regression model, researchers can analyze how these factors collectively affect crop yields. For instance, the model might reveal that soil salinity has a detrimental effect on plant growth, but this effect is moderated by soil texture and organic matter, helping to develop more effective strategies for managing salinity and improving crop performance (Park et al. 2024).

Furthermore, multiple regression can be applied to evaluate the impact of soil management practices on soil health indicators. For example, researchers might use multiple regression to assess how different tillage methods, cover cropping, and organic amendments influence soil microbial activity and nutrient cycling. By including these variables in a regression model, scientists can identify which management practices are most effective in enhancing soil health and supporting sustainable agriculture. This application is particularly valuable for developing evidence-based recommendations for soil management. Overall, the application of multiple regression in soil science provides a robust framework for analyzing complex interactions between soil properties and agricultural outcomes. By integrating various soil characteristics into a single model, researchers can gain deeper insights into how these factors interact and influence soil health, plant growth, and nutrient availability. This understanding allows for more informed decision-making and the development of effective soil management strategies that enhance productivity and sustainability.

Conclusion

Multiple regression is an indispensable tool in soil science and data analysis due to its ability to unravel the intricate relationships between various soil properties and their combined effects on soil health, plant growth, and nutrient availability. By incorporating multiple predictors into a single model, researchers can gain a comprehensive understanding of how factors such as soil texture, chemical properties, and organic carbon interact to influence agricultural outcomes. This method enables scientists to identify key variables, optimize soil management practices, and develop targeted strategies for improving soil fertility and crop productivity. In essence, multiple regression provides a powerful framework for analyzing complex soil data, facilitating evidence-based decision-making and fostering advancements in sustainable agriculture and environmental stewardship.

References

Al Khazaleh M, Karumanchi M, Bellum RR et al (2024) Experimental assessment of geotechnical properties of nano-clay-stabilized soils: advanced sustainable geotechnical solution. Int J Geosynth Ground Eng 10:e8. https://doi.org/10.1007/s40891-023-00517-z

Alzabeebee S, Keawsawasvong S (2024) Robust models to predict the secondary compression index of fine-grained soils using multi objective evolutionary polynomial regression analysis. Model Earth Syst Environ 10:157–165. https://doi.org/10.1007/s40808-023-01778-3

Egbueri JC, Khan MYA (2024) Understanding the geotechnical and geomechanical characteristics of erodible soils: a study incorporating soft computational modeling techniques. Environ Dev Sustain 26:4435–4466. https://doi.org/10.1007/s10668-022-02890-7

Khasawneh MA, Al-Akhrass HI, Rabab'ah SR et al (2024) Prediction of California bearing ratio using soil index properties by regression and machine-learning techniques. Int J Pavement Res Technol 17:306–324. https://doi.org/10.1007/s42947-022-00237-z

Kumar A, Sinha S, Saurav S (2024a) Random forest, CART, and MLR-based predictive model for unconfined compressive strength of cement reinforced clayey soil: a comparative analysis. Asian J Civ Eng 25:2307–2323. https://doi.org/10.1007/s42107-023-00909-6

Kumar A, Sinha S, Saurav S et al (2024b) Prediction of unconfined compressive strength of cement–fly ash stabilized soil using support vector machines. Asian J Civ Eng 25:1149–1161. https://doi.org/10.1007/s42107-023-00833-9

Lipatov A, Belyanova E, Petunina I (2024) A multiple linear regression model to predict the biodegradation rate of soil contaminated with different oil concentrations. Results Nonlinear Anal 7:24–34

Mozaffari H, Moosavi AA, Nematollahi MA (2024) Predicting saturated and near-saturated hydraulic conductivity using artificial neural networks and multiple linear regression in calcareous soils. PLoS One 19:e0296933. https://doi.org/10.1371/journal.pone.0296933

Muhammad A, Danbatta SJ, Muhammad IY et al (2024) Exploring soil radon (Rn) concentrations and their connection to geological and meteorological factors. Environ Sci Pollut Res 31:565–578. https://doi.org/10.1007/s11356-023-31237-6

Park HJ, Baek N, Seo BS et al (2024) Estimation of the electrical conductivity of saturated paste from soil–water extracts of coastal saline paddy soils using random forest and multiple regression models. J Soils Sediments 24:1250–1259. https://doi.org/10.1007/s11368-024-03725-w

Polo-Mendoza R, Duque J, Mašín D (2024) Prediction of California bearing ratio and modified proctor parameters using deep neural networks and multiple linear regression: a case study of granular soils. Case Stud Constr Mater 20:e02800. https://doi.org/10.1016/j.cscm.2023.e02800

Sanjeewani HN, Samarasinghe DP, De Costa WJM (2024) Influence of elevation and the associated variation of climate and vegetation on selected soil properties of tropical rainforests across a wide elevational gradient. Catena 237:107823. https://doi.org/10.1016/j.catena.2024.107823

Shi F, Zhang F, Shen N, Yang M (2024) Quantifying interactions between slope gradient, slope length and rainfall intensity on sheet erosion on steep slopes using Multiple Linear Regression. Sci Total Environ 908:e168090. https://doi.org/10.1016/j.scitotenv.2023.168090

Wu Y, Yang J, Liu X, Lu Y, Lu R (2024) Combined correlation analysis and multilinear regression for strength model of cement-stabilized clayey soils. Int J Geomech 24:e04024190. https://doi.org/10.1061/IJGNAI.GMENG-9579

Zahedifard N, Shahbazi K, Mohammadi MH et al (2024) Soil organic carbon fractions in cultivated calcareous soils. Eurasian Soil Sci 57:780–793. https://doi.org/10.1134/S1064229323603220

Zema DA, Parhizkar M, Plaza-Alvarez PA et al (2024) Using random forest and multiple-regression models to predict changes in surface runoff and soil erosion after prescribed fire. Model Earth Syst Environ 10:1215–1228. https://doi.org/10.1007/s40808-023-01838-8

Principal Component Analysis (PCA)

4

Abstract

Principal Component Analysis (PCA) is a statistical technique used to simplify complex datasets by reducing their dimensionality while preserving as much variance as possible. Principal Component Analysis (PCA) and traditional ANOVA serve different purposes in data analysis, with PCA focusing on reducing dimensionality and uncovering underlying patterns among multiple variables, while ANOVA is used to compare means among different groups. PCA relies on key assumptions including linear relationships between variables, appropriate data scaling, capturing most variance with principal components, and the independence of these components. It handles high-dimensional soil data by simplifying complex datasets into a few principal components, making it easier to identify significant patterns and relationships. In soil science research, PCA is particularly useful for understanding complex interactions between soil properties, improving data visualization, and guiding effective soil management practices. Before applying PCA, researchers should ensure their data meets the assumptions of linearity, scaling, and sufficient variance capture to obtain reliable and interpretable results.

Keywords

Dimensionality reduction · Linear relationships · Principal component analysis (PCA) · Soil Science · Variance

Questions Covered in the Chapter

(a) What is the primary difference between PCA and conventional ANOVA?
(b) What are the main assumptions required for PCA?
(c) How does PCA handle high-dimensional soil data?

T. Souza, *Advanced Statistical Analysis for Soil Scientists*,
https://doi.org/10.1007/978-3-031-88161-9_4

(d) Why is PCA particularly useful in soil science research?
(e) What should researchers check for before applying PCA to ensure reliable results?

Introduction

Principal Component Analysis (PCA) is a statistical technique used to simplify complex datasets by reducing their dimensionality while preserving as much variance as possible. It achieves this by transforming the original variables into a new set of uncorrelated variables called principal components. These components are ordered such that the first few retain most of the variation present in the original dataset. By reducing the number of dimensions, PCA helps in visualizing data, identifying patterns, and making data easier to analyze, particularly when dealing with large and multivariate datasets (Callén et al. 2024).

PCA relies on several key assumptions. Firstly, it assumes linearity, meaning that the relationships between variables are linear and can be represented in a lower-dimensional space. Secondly, PCA assumes that the data has been scaled or standardized, particularly when variables have different units or ranges, to ensure that all variables contribute equally to the analysis. Lastly, PCA assumes that the principal components capture the most significant variance in the data, with the first few components accounting for the majority of the variability, allowing for effective dimensionality reduction (Costa et al. 2024).

As an advanced statistical tool, PCA offers several advantages. It reduces the complexity of high-dimensional data by identifying the most important features, enabling researchers to focus on the most relevant information. This reduction in dimensionality also aids in avoiding overfitting in predictive models by eliminating redundant or less significant variables. Additionally, PCA can improve data visualization, making it easier to detect patterns, clusters, and outliers in large datasets. These benefits make PCA a powerful tool for exploratory data analysis and a foundational technique in fields like machine learning and data science (Juhos et al. 2024).

In soil science, PCA plays a crucial role in analyzing and interpreting complex datasets derived from soil properties, composition, and environmental variables. By reducing the dimensionality of soil data, PCA helps in identifying the most influential factors affecting soil health, fertility, and ecosystem functions. For instance, PCA can be used to discern the key soil properties that influence crop yield, to categorize different soil types based on their chemical and physical characteristics, or to study the impact of land use changes on soil biodiversity. As a result, PCA enables soil scientists to make more informed decisions in agriculture, environmental management, and conservation efforts (Kulmány et al. 2024).

Main Differences Between PCA and Conventional ANOVA

The primary difference between Principal Component Analysis (PCA) and Analysis of Variance (ANOVA) lies in their objectives and the type of data they analyze. PCA is a dimensionality reduction technique used to simplify complex datasets by transforming correlated variables into a set of uncorrelated principal components. It focuses on identifying patterns and capturing the maximum variance within the data, making it suitable for exploratory data analysis. In contrast, ANOVA is a statistical method used to compare the means of different groups or treatments to determine if there are statistically significant differences between them. ANOVA is primarily used for hypothesis testing in experimental designs (Menge et al. 2024).

PCA is concerned with understanding the underlying structure of multivariate data. It does so by identifying the directions (principal components) in which the data varies the most. In soil science, for instance, when analyzing a dataset containing multiple soil chemical properties such as pH, organic matter, nitrogen content, and phosphorus levels, PCA could be used to identify which of these properties contribute the most to the variability in the dataset. This helps in simplifying the data by reducing it to a few principal components that represent the major patterns without losing significant information (Miao et al. 2024).

ANOVA, on the other hand, is focused on testing differences between group means. For example, if a soil scientist wants to compare the mean nitrogen content of soil samples taken from different land management practices (e.g., conventional farming, organic farming, and no-till farming), ANOVA would be the appropriate method. It would determine whether the observed differences in nitrogen content across the different practices are statistically significant or if they could have occurred by random chance. ANOVA provides insight into whether specific treatments or conditions have a significant impact on soil properties (Roy and Chintalacheruvu 2024).

In the field of soil science, PCA could be applied to a comprehensive dataset of various soil chemical properties to identify the most influential factors driving soil variability across a region. For example, if a study is conducted to assess soil fertility across different land types, PCA might reveal that variables like organic matter and phosphorus are the primary contributors to soil fertility variations. ANOVA, in contrast, would be useful when comparing the mean values of specific soil properties (like pH or potassium levels) across different experimental treatments or land use categories to assess the impact of these factors on soil health. While PCA provides a broad overview of the data structure, ANOVA is more focused on testing specific hypotheses related to differences in soil properties across groups or treatments (Reza et al. 2024).

Key Concepts of PCA

When studying soil, scientists often measure many properties like pH, organic matter, nitrogen, phosphorus, and more. These properties give us a detailed picture of the soil, but when there are many measurements, it can become challenging to analyze and understand the data. Principal Component Analysis (PCA) helps by reducing the number of variables we need to consider while still keeping the important information. Imagine you have a big puzzle with many pieces—PCA helps you focus on just a few key pieces that tell most of the story (Ali et al. 2024).

In soil science, PCA takes all those different measurements, such as chemical properties, and combines them into a few new variables called principal components. These components are like summaries of the original data. For example, the first principal component might represent a combination of soil properties that are most strongly related to plant nutrition, like nitrogen and phosphorus levels. By focusing on these components, we can better understand which soil properties are most important for plant growth without getting lost in too many details (Candan et al. 2024).

One of the most useful aspects of PCA is its ability to reveal patterns. For instance, if you're studying the nutrition levels in plants across different soil types, PCA can help identify which soil properties are most closely linked to healthy plant growth. Perhaps you'll find that certain soil characteristics, like high organic matter and balanced pH, are consistently associated with better nutrition in plants. By highlighting these patterns, PCA helps you see connections that might not be obvious when looking at each soil property separately (Ji et al. 2024).

For young soil scientists, understanding PCA is like gaining a new tool for making better decisions about soil management and plant nutrition. If you know that certain soil properties are critical for plant health, you can focus your efforts on managing those properties more effectively. For example, if PCA shows that phosphorus levels are a key factor in crop yield, you can prioritize phosphorus management in your soil treatment plans. This not only makes your work more efficient but also leads to better outcomes in agriculture and environmental conservation (Hammoumi et al. 2024).

To Summarize

1. Understanding Dimensionality in a robust database
2. Simplifying complex data – Reducing *n*-dimensionality
3. Finding Main Patterns in the database
4. Making Better Decisions by interpreting the research results

Main Assumptions Required for PCA

The first major assumption of Principal Component Analysis (PCA) is that the relationships between variables are linear. This means that PCA assumes that the variation in the data can be captured by straight-line relationships between variables. For instance, when analyzing the dynamics of phosphorus (P) in acidic soils, PCA assumes that changes in P availability are linearly related to other soil properties like pH, organic matter, or aluminum content. If the relationships are non-linear, PCA may not fully capture the complexities of how these factors interact in the soil (Gharbia et al. 2024).

Another key assumption is that the data has been appropriately scaled or standardized, especially when the variables have different units or ranges. In the context of acidic soils, this might involve standardizing variables like pH, P content, and organic matter, so they all contribute equally to the PCA. Without scaling, variables with larger numerical ranges could dominate the analysis, leading to misleading results. For example, if pH (usually a small number) is not scaled, its influence on the principal components could be underestimated compared to the P concentration (which might have a wider range of values) (Liu et al. 2024).

PCA assumes that the principal components that capture the most variance in the data are the most important for understanding the underlying structure. In soil science, when studying P dynamics in acidic soils, PCA would prioritize the components that explain the most variation in properties like P availability, pH, and organic matter. This assumption implies that the most significant soil characteristics affecting P dynamics will be reflected in these components. If the most variance in the data is not related to the aspects of interest (like P availability), PCA might not effectively highlight the important factors (Mikha et al. 2024).

Finally, PCA assumes that the principal components are uncorrelated or independent of each other. This means that each component represents a distinct pattern of variation in the data. In the context of acidic soils, the first principal component might reflect the influence of pH on P availability, while the second might represent the impact of organic matter. These components should not overlap in the variation they capture, allowing for a clearer understanding of the different factors affecting P dynamics. If the components are not truly independent, the interpretation of the results could become more complex and less meaningful (Reis and Dindaroğlu 2024).

Steps to Conduct PCA

The first step in conducting Principal Component Analysis (PCA) is gathering and preparing your data. For soil scientists studying saline-sodic soils, this might involve collecting data on various soil physical properties such as soil texture, bulk density, porosity, and soil moisture content. Additionally, you might measure plant yield in these soils to understand how soil properties influence crop production. Once you have your data, it's crucial to standardize or normalize the variables, especially if they are measured in different units. For example, soil moisture might be in percentage, while bulk density is in grams per cubic centimeter. Standardization ensures that each variable contributes equally to the PCA (Moghim et al. 2024).

After preparing the data, the next step is to calculate the covariance matrix, which shows how pairs of variables in your dataset vary together. In the context of saline-sodic soils, the covariance matrix would help you understand the relationships between soil properties like bulk density and porosity. For example, if high soil salinity often corresponds with low porosity, this relationship will be reflected in the covariance matrix. Understanding these relationships is key to determining how the different soil physical properties are interconnected and how they might influence plant yield (Reis and Dindaroğlu 2024).

Once the covariance matrix is calculated, the next step is to extract the principal components. These components are new variables that represent the directions in which your data varies the most. In the case of soil physical properties in saline-sodic soils, the first principal component might capture the most significant variation, perhaps reflecting the combined effects of soil texture and salinity on plant yield. Each subsequent component captures less variance but represents other patterns in the data, such as the influence of soil moisture on crop performance. These components help simplify the complex dataset into a more manageable form (Tibane and Mamba 2024).

The final step in PCA is interpreting the principal components and using them to inform your research or management decisions. For young soil scientists, this involves understanding what each principal component represents. For example, if the first principal component shows that soil salinity and bulk density strongly influence plant yield, you might focus on managing these factors to improve crop performance. PCA allows you to identify which soil properties are most critical in saline-sodic soils and helps prioritize interventions that could enhance plant yield. By focusing on the key components, you can make more informed decisions about soil management and crop production strategies (Wei et al. 2024).

To Summarize

Step 1: Data Collection and Preparation

Step 2: Calculating the Covariance Matrix

Step 3: Extracting Principal Components

Step 4: Interpreting the Results and Making Decisions

Handling High-Dimensional Soil Data with PCA

Principal Component Analysis (PCA) is particularly useful for handling high-dimensional data, where the number of variables is large compared to the number of observations. In soil science, high-dimensional data might include measurements of all essential micronutrients, such as zinc (Zn), iron (Fe), manganese (Mn), copper (Cu), boron (B), molybdenum (Mo), and chlorine (Cl). Analyzing each micronutrient individually can be overwhelming and may obscure the relationships between them. PCA simplifies this complexity by reducing the dimensionality of the data, allowing scientists to focus on the most important patterns and interactions among these nutrients (Yi et al. 2024).

PCA reduces redundancy in high-dimensional data by identifying and removing correlations between variables. In the case of soil micronutrients, some of these elements might be closely related, such as Zn and Fe, which often behave similarly in soil environments. PCA combines these correlated variables into a smaller number of principal components. For example, the first principal component might represent a combination of Zn, Fe, and Mn levels, capturing the most significant variance in the dataset. By reducing the data to a few components, PCA helps soil scientists identify the key patterns that affect soil fertility and plant health (Zeng et al. 2024).

One of the challenges of high-dimensional data is that it's difficult to visualize and interpret. PCA addresses this by projecting the high-dimensional data into a lower-dimensional space, typically two or three dimensions, which can be easily visualized. For instance, when studying the impact of micronutrient levels on crop yield, PCA can create a scatter plot of the first two or three principal components. This plot might reveal clusters or trends, such as specific micronutrient profiles associated with high or low yields. By providing a clear visual representation, PCA makes it easier for soil scientists to interpret complex relationships in the data (Yoon et al. 2024).

By concentrating the variance into a few principal components, PCA allows soil scientists to focus on the most influential micronutrients that drive soil fertility and crop production. For example, after performing PCA on a dataset of all essential micronutrients, you might find that Zn and B are the primary factors influencing plant growth in a specific soil type. This insight enables you to prioritize the management of these nutrients in your soil treatment plans. Instead of treating all micronutrients equally, PCA guides you to target the ones that have the most significant impact on soil health and crop yield, leading to more effective and efficient soil management practices (Zhang et al. 2024).

Why PCA Is Particularly Useful in Soil Science Research?

Soil science often involves studying numerous interconnected variables, such as soil properties, plant health, and microbial communities. Principal Component Analysis (PCA) is particularly useful in soil science because it simplifies this complexity by reducing the number of variables that need to be analyzed. For example, when examining soil microbial communities, researchers might measure various ecological indices like species richness, diversity, evenness, and abundance. These indices can be overwhelming to analyze all at once. PCA helps by transforming these multiple variables into a few principal components that capture the most significant patterns in the data, making it easier to understand and interpret (Tibane and Mamba 2024).

PCA is valuable for uncovering hidden patterns and relationships in complex datasets that might not be immediately apparent through traditional analysis. In the context of soil microbial communities, PCA can reveal how different environmental factors or soil properties influence microbial diversity and function. For example, PCA might show that the first principal component correlates strongly with soil organic matter and moisture content, suggesting that these factors play a critical role in shaping microbial communities. By highlighting these underlying patterns, PCA allows soil scientists to gain deeper insights into the ecological dynamics of soil microbes (Zhu et al. 2024).

Interpreting high-dimensional data visually is challenging, but PCA addresses this by reducing the dimensionality of the data, making it easier to create meaningful visualizations. For instance, when studying the impact of agricultural practices on soil microbial communities, PCA can project the data onto two or three dimensions, allowing researchers to plot the microbial community structure and ecological indices in a more interpretable way. This might reveal distinct clusters of microbial communities associated with different farming practices, such as organic versus conventional farming, helping scientists better understand the ecological impacts of these practices (Zhou et al. 2024).

PCA is also instrumental in guiding targeted soil management strategies by identifying the most influential factors affecting soil health and microbial ecology. For example, if PCA analysis of soil microbial communities shows that soil pH and organic matter are the primary drivers of microbial diversity, soil scientists can focus their efforts on managing these properties to enhance soil health. By concentrating on the most impactful variables, PCA helps researchers and practitioners develop more effective and efficient soil management practices, ultimately leading to improved soil sustainability and productivity (Ali et al. 2024).

Practical Application in Agroforestry vs. Conventional Farming Systems

The practical application of Principal Component Analysis (PCA) in comparing agroforestry with conventional farming systems provides a powerful way to analyze and interpret complex agricultural data. In these two distinct farming systems, numerous factors such as soil properties, crop yields, biodiversity, and environmental impacts come into play. PCA allows researchers to distill this vast amount of data into a few key components, making it easier to compare the overall effects and outcomes of each system. For instance, in agroforestry systems, where multiple crops and trees are grown together, the data may include variables related to soil fertility, water retention, and species diversity. In contrast, conventional farming might focus more on monoculture yields and chemical inputs. PCA can combine these diverse variables into principal components that summarize the main differences and similarities between the systems (Callén et al. 2024).

By applying PCA, researchers can identify which factors are most influential in differentiating agroforestry from conventional farming. For example, PCA might reveal that the first principal component is heavily influenced by variables such as soil organic matter, biodiversity indices, and nutrient cycling efficiency, which are typically higher in agroforestry systems. This insight allows scientists to pinpoint the ecological benefits of agroforestry, such as improved soil health and greater resilience to environmental changes, compared to conventional farming. Similarly, another principal component might highlight the economic aspects, such as crop yield stability and input costs, providing a balanced view of the trade-offs between the two systems (Gharbia et al. 2024).

The use of PCA in this context also aids in visualizing complex relationships within the data, making it easier to communicate findings to stakeholders, including farmers, policymakers, and agricultural researchers. For example, a PCA plot could visually represent how different farming practices cluster together based on their impact on soil health or biodiversity. In agroforestry systems, where a diverse range of species and practices are integrated, PCA can help demonstrate how these practices collectively contribute to sustainable farming outcomes. Conversely, the PCA might show how conventional farming practices cluster around higher chemical input use and lower biodiversity, highlighting areas where improvements could be made (Ji et al. 2024).

Ultimately, PCA serves as a valuable tool for guiding decision-making in agricultural systems. By revealing the most critical factors that distinguish agroforestry from conventional farming, it helps stakeholders make informed choices about which practices to adopt or promote. For instance, if PCA indicates that agroforestry significantly enhances soil health and biodiversity without sacrificing yield, policymakers might be encouraged to support agroforestry initiatives. On the other hand, if conventional farming practices show clear advantages in certain economic aspects, these insights can guide strategies to optimize these systems. In this way, PCA not only enhances our understanding of the complexities within different farming systems but also provides a practical foundation for developing more sustainable and productive agricultural landscapes (Menge et al. 2024).

Applying PCA in Soil Science

When working with complex datasets in soil science, applying multiple Principal Component Analyses (PCA) can be a powerful approach to understanding different aspects of soil properties and management. For example, in clay soils, researchers might be interested in studying soil physical properties, soil quality, and cation exchange capacity (CEC). By applying PCA separately to each of these categories, scientists can uncover the key factors that influence each aspect of soil behavior. For instance, one PCA might be used to analyze soil physical properties like bulk density, porosity, and texture, while another PCA focuses on soil quality indicators such as organic matter, nutrient content, and microbial activity (Juhos et al. 2024; Liu et al. 2024).

By conducting multiple PCA, young scientists can compare and contrast the principal components from each analysis to gain a more comprehensive understanding of how different soil properties interact. In clay soils, for example, the PCA of physical properties might reveal that porosity and bulk density are the primary factors influencing water retention. Meanwhile, a separate PCA on soil quality might highlight the importance of organic matter and nutrient availability for supporting plant growth. When these results are considered together, they provide a holistic view of soil health, showing how physical structure and chemical composition both contribute to the overall quality of the soil (Reis and Dindaroğlu 2024).

Multiple PCA can also help identify the relationships between different soil properties and management practices. For example, when studying the cation exchange capacity (CEC) in clay soils, a PCA might reveal which soil amendments or management practices have the most significant impact on CEC. If a PCA shows that certain practices, like the addition of organic matter, strongly correlate with increased CEC, scientists can use this information to guide soil management strategies. Similarly, another PCA might show how these practices influence physical properties like aggregation or compaction, helping to balance different aspects of soil management (Mikha et al. 2024).

Ultimately, applying multiple PCA in soil science allows researchers to break down complex datasets into more manageable components, making it easier to understand and improve soil management practices. By analyzing different soil properties separately and then integrating the results, young scientists can develop targeted strategies that address the specific needs of clay soils. Whether the goal is to enhance soil structure, boost nutrient retention, or improve overall soil quality, multiple PCA provides a flexible and effective tool for guiding decision-making and advancing sustainable soil management practices (Reza et al. 2024).

Software for PCA

Several software tools are available for conducting Principal Component Analysis (PCA), each offering various features suited to different needs and levels of expertise. Here's a brief overview of some popular software options:

1. **R and RStudio:** R is a powerful statistical programming language widely used for performing PCA and other statistical analyses. With packages like stats (which includes the prcomp and princomp functions for PCA) and FactoMineR, R provides comprehensive tools for performing and visualizing PCA. RStudio, an integrated development environment (IDE) for R, enhances the user experience with a user-friendly interface and additional features for data visualization and analysis. R's extensive ecosystem of packages allows for customization and advanced analysis, making it a preferred choice for many researchers and data scientists.
2. **Python** (with libraries like scikit-learn and pandas): Python, with libraries such as scikit-learn and pandas, is another popular choice for PCA. The scikit-learn library provides the PCA class for performing principal component analysis, and pandas helps in data manipulation and preprocessing. Python's visualization libraries, like matplotlib and seaborn, can be used to create plots of the PCA results, such as biplots and scree plots. Python's versatility and readability make

it a great option for both beginners and advanced users working on PCA and other data analysis tasks.

3. **MATLAB**: It is a high-level programming environment that offers built-in functions for PCA through its Statistics and Machine Learning Toolbox. The pca function in MATLAB provides an easy way to perform PCA and analyze the results. MATLAB is known for its powerful matrix operations and visualization capabilities, making it suitable for handling large datasets and complex analyses. It's widely used in academic and industrial settings for data analysis and modeling.
4. **SPSS (Statistical Package for the Social Sciences)**: SPSS is a user-friendly software package widely used in social sciences and other fields for statistical analysis, including PCA. It provides a graphical interface that simplifies the process of conducting PCA and interpreting the results. SPSS is useful for users who prefer a point-and-click approach rather than programming. It includes various options for output visualization, such as component plots and biplots, which facilitate the interpretation of PCA results.
5. **Excel with Add-ins**: Microsoft Excel, often used for basic data analysis, can perform PCA with the help of add-ins like the XLSTAT or the Analyse-it add-in. These add-ins provide additional statistical capabilities, including PCA, and integrate seamlessly with Excel's data handling and visualization features. While not as advanced as other dedicated statistical software, these add-ins offer a convenient option for users familiar with Excel who need to perform PCA without switching to a different platform.

Example Application in Soil Science

In soil science, Principal Component Analysis (PCA) can be effectively applied to understand and manage soil health by analyzing complex datasets with multiple soil properties. For example, researchers studying soil fertility might measure various attributes such as soil pH, organic matter content, nutrient levels, and moisture content across different agricultural fields. By applying PCA to this data, scientists can reduce the dimensionality and identify the key factors that contribute most significantly to soil fertility. This allows them to simplify the analysis and focus on the primary variables influencing soil quality, making it easier to interpret the results and apply targeted soil management practices.

PCA can also be used to compare soil properties across different land management practices, such as organic versus conventional farming. In this scenario, data on soil properties like texture, bulk density, and nutrient levels are collected from fields managed using different approaches. Applying PCA helps to reveal how these management practices affect soil characteristics by summarizing the main differences and similarities between the systems. For instance, PCA might show that organic farming practices are associated with higher organic matter content and improved soil structure, while conventional farming might be linked to higher nutrient levels but lower soil health indicators. This insight helps in evaluating the effectiveness of various farming practices and guiding recommendations for sustainable land management.

When dealing with soil pollution or contamination, PCA can assist in identifying the sources and impacts of contaminants. Researchers might collect soil samples from different locations affected by various types of pollution, such as heavy metals or pesticides. By applying PCA, they can uncover patterns related to the distribution of contaminants and their interactions with soil properties. For example, PCA might reveal that high levels of specific heavy metals correlate with lower soil pH and reduced microbial activity. This information is crucial for assessing the extent of contamination and developing strategies for soil remediation and pollution control.

In studies of soil microbial communities, PCA can be used to analyze the relationships between soil properties and microbial diversity. Researchers might measure soil characteristics such as nutrient availability, moisture, and temperature, along with microbial indices like species richness and abundance. Applying PCA helps to identify which soil properties are most strongly associated with variations in microbial communities. For instance, PCA might show that soil moisture and organic matter content are key factors influencing microbial diversity and activity. This understanding helps in managing soil conditions to support healthy microbial populations, which are essential for soil fertility and ecosystem functioning.

PCA can also be applied to understand the impact of soil amendments on soil properties and plant growth. Researchers might compare the effects of different soil amendments, such as compost, lime, or chemical fertilizers, on soil attributes and crop yields. By using PCA, they can evaluate how these amendments alter soil characteristics and identify the most effective treatments for improving soil health. For example, PCA might show that compost and lime significantly increase soil organic matter and pH, leading to better plant growth compared to chemical fertilizers, which might only affect nutrient levels. This insight informs decisions on the best soil amendments to use for optimizing crop production and soil health.

Finally, PCA can be instrumental in assessing soil quality and sustainability over time. By collecting longitudinal data on soil properties, such as nutrient levels, pH, and organic matter, researchers can apply PCA to track changes and identify trends. This analysis helps in evaluating the long-term effects of different soil management practices and environmental conditions on soil quality. For instance, PCA might reveal that certain practices, like reduced tillage or cover cropping, lead to improvements in soil structure and fertility over time, providing evidence for their benefits in sustainable soil management. This approach helps in making informed decisions to maintain and enhance soil quality for future agricultural and environmental needs.

Conclusion

Principal Component Analysis (PCA) is a vital statistical tool in soil science and data analysis, offering a robust method for managing and interpreting complex datasets with multiple variables. By reducing dimensionality and identifying key principal components that capture the most significant variance in the data, PCA simplifies the analysis of intricate relationships among soil properties, plant health, and environmental factors. In soil science, PCA facilitates the understanding of how various soil attributes—such as pH, organic matter, nutrient levels, and microbial activity—interact and influence one another, enabling more effective soil management and improvement strategies. Its ability to distill complex data into actionable insights makes PCA indispensable for researchers and practitioners aiming to enhance soil health, optimize agricultural practices, and address environmental challenges. Through its capacity to reveal underlying patterns and correlations, PCA not only streamlines data analysis but also supports informed decision-making, ultimately contributing to sustainable and productive soil management practices.

References

Ali Z, Waheed H, Malik RN (2024) Deciphering integrated soil potential ecological and human health risks attributable to industrial effluents irrigation and wheat consumption. Environ Sci Pollut Res. https://doi.org/10.1007/s11356-024-34472-7

Callén MS, Martínez I, Grasa G et al (2024) Principal component analysis and partial least square regression models to understand sorption-enhanced biomass gasification. Biomass Convers Biorefin 14:2091–2111. https://doi.org/10.1007/s13399-022-02496-z

Candan F, Markushin Y, Ozbay G (2024) Nanoparticle uptake and bioaccumulation in *Pisum sativum* L. (Green Pea) analyzed via dark-field microscopy, infrared spectroscopy, and principal component analysis combined with machine learning. Agronomy 14:e1473. https://doi.org/10.3390/agronomy14071473

Costa AG, Silva CJD, Fonseca CSD et al (2024) Multivariate analysis applied to evaluate the spatial variability of a soil subjected to conservation management. Rev Ciênc Agron 55:e20228656. https://doi.org/10.5935/1806-6690.20240041

Gharbia AS, Zákányi B, Tóth M (2024) Impact of sand media continuous drying and rewetting cyclic on nutrients transformation performance from reclaimed wastewater effluent at soil aquifer treatment. Sci Rep 14:8065. https://doi.org/10.1038/s41598-024-58787-0

Hammoumi D, Al-Aizari HS, Alaraidh IA et al (2024) Seasonal variations and assessment of surface water quality using water quality index (WQI) and principal component analysis (PCA): a case study. Sustainability 16:5644. https://doi.org/10.3390/su16135644

Ji L, Ma K, Xie TN et al (2024) Evaluation of heavy metal distribution characteristics and ecological risk of soil of vegetable land for Hong Kong in Ningxia. Huan Jing Ke Xue 45:3512–3522. https://doi.org/10.13227/j.hjkx.202307048

Juhos K, Nugroho PA, Jakab G et al (2024) A comprehensive analysis of soil health indicators in a long-term conservation tillage experiment. Soil Use Manag 40:e12942. https://doi.org/10.1111/sum.12942

Kulmány IM, Bede L, Stencinger D et al (2024) Challenges in mapping soil variability using apparent soil electrical conductivity under heterogeneous topographic conditions. Agronomy 14:1161. https://doi.org/10.3390/agronomy14061161

Liu H, Anwar S, Fang L et al (2024) Source apportionment of agricultural soil heavy metals based on PMF model and multivariate statistical analysis. Environ Forensic 25:40–48. https://doi.org/10.1080/15275922.2022.2047838

Menge DM, Musila RN, Kagito S et al (2024) Using principal component analysis to assess soil chemical properties in the Mwea irrigation Scheme, Kenya: implications for rice agronomic management. Int J Plant Soil Sci 36:106–126. https://doi.org/10.9734/ijpss/2024/v36i54508

Miao Y, Wang H, Huang X et al (2024) Soil organic carbon prediction based on vis–NIR spectral classification data using GWPCA–FCM algorithm. Sensors 24:4930. https://doi.org/10.3390/s24154930

Mikha MM, Green TR, Untiedt TJ, Hergret GW (2024) Land management affects soil structural stability: multi-index principal component analyses of treatment interactions. Soil Tillage Res 235:105890. https://doi.org/10.1016/j.still.2023.105890

Moghim MF, Karimi A, Bagheri-Bodaghabadi M et al (2024) Evaluation of soil quality and land suitability in different management systems. Arch Agron Soil Sci 70:1–19. https://doi.org/10.1080/03650340.2023.2298851

Reis A, Dindaroğlu T (2024) Evaluating dynamic soil quality by the soil management assessment framework (SMAF) in the watershed scale in a semi-arid Mediterranean ecosystem in Turkey. Geoderma Reg 38:e00829. https://doi.org/10.1016/j.geodrs.2024.e00829

Reza SK, Sharma GK, Alam NM et al (2024) Assessing soil quality under different land-uses through constructing minimum datasets from soil profiles in a fragile ecosystem of north-eastern region of India. Commun Soil Sci Plant Anal 55:1629–1643. https://doi.org/10.1080/00103624.2024.2323075

Roy S, Chintalacheruvu MR (2024) Enhanced morphometric analysis for soil erosion susceptibility mapping in the Godavari river basin, India: leveraging Google Earth Engine and principal component analysis. ISH J Hydraul Eng 30:228–244. https://doi.org/10.1080/09715010.2023.2292280

Tibane LV, Mamba D (2024) Environmental risk assessment, principal component analysis, tracking the source of toxic heavy metals of solid gold mine waste tailings, South Africa. Environ Forensic 25:254–270. https://doi.org/10.1080/15275922.2023.2172478

Wei W, Yan P, Zhou L et al (2024) A comprehensive drought index based on spatial principal component analysis and its application in northern China. Environ Monit Assess 196:193. https://doi.org/10.1007/s10661-024-12366-y

Yi X, Lu Y, He J et al (2024) Changes in soil erosion resistance after farmland abandonment in the Karst Rocky desertification area of Southwest China. J Soil Sci Plant Nutr 24:3519–3536. https://doi.org/10.1007/s42729-024-01773-2

Yoon JH, Lee CG, Park BJ et al (2024) Combining soil immobilization and dressing techniques for sustaining the health of metal-contaminated arable soils. Sustainability 16:3227. https://doi.org/10.3390/su16083227

Zeng W, Zhang Y, Luo W et al (2024) Soil-microbial interactions in rice-loach-shrimp integrated farming: multivariate analysis of ecological intensification. Water 16:2083. https://doi.org/10.3390/w16152083

Zhang J, Dyck M, Quideau SA, Norris CE (2024) Assessment of soil health and identification of key soil health indicators for five long-term crop rotations with varying fertility management. Geoderma 443:116836. https://doi.org/10.1016/j.geoderma.2024.116836

Zhou Y, Feng F, Sun J, Shan Y, Su W, Shang W, Li Y (2024) Distribution and source analysis of soil toxic organic compounds of coal-electricity production base in arid and semi-arid region of China. J Hazard Mater 135317. https://doi.org/10.1016/j.jhazmat.2024.135317

Zhu Z, Chen J, Hu H et al (2024) Soil quality evaluation of different land use modes in small watersheds in the hilly region of southern Jiangsu. Ecol Indic 160:111895. https://doi.org/10.1016/j.ecolind.2024.111895

Redundancy Analysis (RDA) 5

Abstract

Redundancy Analysis (RDA) is a multivariate statistical technique that extends principal component analysis (PCA) by incorporating explanatory variables. RDA and PCA differ primarily in their focus and application. While PCA is used for dimensionality reduction by identifying principal components that explain the most variance in the data, RDA is designed to explore the relationships between a set of response variables and explanatory variables, providing insights into how environmental or management factors influence soil properties. The main assumptions required for RDA include linear relationships between variables, normality, homogeneity of variances, and the independence of observations. RDA handles high-dimensional soil data by reducing complexity through ordination, which allows for simultaneous analysis of multiple response variables and their relationships with explanatory variables. This makes RDA particularly useful in soil science research as it helps elucidate complex interactions among soil properties and management practices, guiding sustainable soil management strategies. Before applying RDA, researchers should check for assumptions such as linearity, data distribution, and multicollinearity, and ensure proper data scaling and transformation to achieve reliable and interpretable results.

Keywords

Environmental factors · Explanatory variables · Multivariate statistical technique · Redundancy analysis (RDA) · Soil properties

Questions Covered in the Chapter

(a) What is the primary difference between RDA and PCA?
(b) What are the main assumptions required for RDA?

T. Souza, *Advanced Statistical Analysis for Soil Scientists*,
https://doi.org/10.1007/978-3-031-88161-9_5

(c) How does RDA handle high-dimensional soil data?
(d) Why is RDA particularly useful in soil science research?
(e) What should researchers check for before applying RDA to ensure reliable results?

Introduction

Redundancy Analysis (RDA) is a multivariate statistical technique that extends principal component analysis (PCA) by incorporating explanatory variables. Unlike PCA, which focuses solely on reducing the dimensionality of response variables (such as species abundance or soil properties), RDA simultaneously considers the relationship between these response variables and a set of explanatory variables (such as environmental factors). This method allows researchers to explore the variation in the response data that can be explained by the explanatory variables, making RDA a powerful tool in ecological and environmental studies (Zhu et al. 2024).

RDA is based on several key assumptions. First, it assumes linear relationships between the response and explanatory variables, meaning that the changes in response variables can be well-explained by a linear combination of the explanatory variables. Second, it assumes that the data are multivariate normal, though in practice, RDA can be relatively robust to deviations from this assumption. Finally, like other ordination techniques, RDA assumes that the datasets being analyzed are free from multicollinearity, which could otherwise distort the analysis by inflating the variance explained by correlated variables (Zhao et al. 2024a).

One of the primary advantages of RDA as an advanced statistical analysis method is its ability to handle complex datasets that involve multiple response and explanatory variables. This makes it especially useful in environmental and ecological research, where datasets are often multidimensional. RDA provides a clear visualization of how much of the variation in response variables can be attributed to specific environmental gradients or experimental treatments. This makes it an invaluable tool for identifying key environmental factors that influence ecological communities or other complex systems (Shao et al. 2024).

In soil science, RDA is particularly important because it allows researchers to understand the relationships between soil properties and various biotic and abiotic factors. For instance, when studying soil microbial communities, RDA can be used to link variations in microbial composition to soil chemical properties, such as pH, nutrient levels, or organic matter content. By doing so, researchers can identify which soil properties are most influential in shaping the microbial community structure, providing insights into soil health and fertility (Peng et al. 2024).

Additionally, RDA can be applied to study the relationship between soil physical properties, such as texture, moisture content, and bulk density, and various environmental or management factors. For example, RDA could be used to explore how different land-use practices impact soil structure, with implications for soil conservation and sustainable agriculture. By identifying the main drivers of soil physical properties, researchers and land managers can develop strategies to maintain or improve soil quality (Li et al. 2024a).

Overall, RDA is a versatile and powerful statistical tool in soil science, offering the ability to dissect complex relationships between soil properties and environmental factors. Its applications extend from understanding the factors driving microbial diversity in soils to informing soil management practices aimed at preserving or enhancing soil function. The insights gained from RDA analyses are crucial for advancing our understanding of soil ecosystems and developing strategies for sustainable land management (Mu et al. 2024).

Main Differences Between RDA and PCA

The primary difference between Redundancy Analysis (RDA) and Principal Component Analysis (PCA) lies in their approach to handling explanatory variables and the objectives they serve in data analysis. PCA is an ordination technique that focuses solely on summarizing the variation within a set of response variables without considering any external factors or explanatory variables. It achieves this by finding linear combinations of the response variables that capture the maximum variance in the dataset. PCA is purely descriptive and is mainly used to reduce the dimensionality of a dataset, making it easier to visualize and interpret patterns in the response variables (Jia et al. 2024a).

In contrast, RDA extends the concept of PCA by incorporating explanatory variables into the analysis. This means that RDA not only looks at the variation within the response variables but also examines how this variation is related to a set of explanatory variables. RDA achieves this by constraining the ordination of response variables based on the linear relationships with the explanatory variables. As a result, RDA provides a more targeted analysis by identifying how much of the variation in the response variables can be explained by the explanatory variables. This makes RDA particularly useful for studies where understanding the influence of specific environmental or experimental factors on a set of response variables is crucial (Chen et al. 2024a).

To illustrate the difference, consider a study on plant nutrition where researchers are investigating the effects of varying levels of copper (Cu) and zinc (Zn) on plant growth and nutrient uptake. If the researchers were to use PCA, they would analyze the variation in plant growth parameters (such as biomass, leaf chlorophyll content, and nutrient concentrations) without considering the Cu and Zn levels. PCA would reveal patterns and groupings within the plant growth data, helping to identify which growth parameters are most closely related to each other. However, PCA would not provide any direct insights into how the Cu and Zn levels influence these growth parameters (Bi et al. 2024).

On the other hand, if the researchers used RDA, they could explicitly include the Cu and Zn levels as explanatory variables in the analysis. RDA would then allow them to assess how much of the variation in plant growth parameters can be attributed to the differences in Cu and Zn levels. For example, RDA might reveal that a significant portion of the variation in leaf chlorophyll content is explained by Zn levels, while biomass variation is more strongly influenced by Cu levels. This targeted analysis would provide valuable insights into the specific roles of Cu and Zn in plant nutrition, which could inform fertilizer management practices (Chang et al. 2024).

Another key difference is in the interpretation of the results. In PCA, the principal components are linear combinations of all the response variables, and these components are interpreted as new, uncorrelated variables that summarize the original dataset. The axes in a PCA plot represent these principal components, with the first axis capturing the most variance and subsequent axes capturing progressively less. In RDA, the axes represent the variation in response variables that is explained by the explanatory variables. The RDA axes are interpreted in relation to the explanatory variables, allowing researchers to directly assess the impact of these variables on the response data. This direct relationship between the response and explanatory variables makes RDA a more powerful tool for hypothesis testing and ecological inference (Chen et al. 2024b).

Ultimately, the choice between PCA and RDA depends on the research objectives. PCA is more appropriate when the goal is to explore and summarize the structure of response variables without making assumptions about external influences. RDA, on the other hand, is ideal when researchers are interested in understanding how specific factors or gradients influence the response variables. In the context of plant nutrition studies, RDA would be particularly useful for disentangling the effects of Cu and Zn on various growth and nutrient parameters, providing a clearer picture of how these elements interact with plant physiology.

Key Concepts of RDA

Redundancy Analysis (RDA) is an advanced multivariate statistical method that helps researchers explore the relationships between multiple response variables and explanatory variables simultaneously. Unlike some other statistical methods, RDA focuses on understanding how much of the variation in a set of response variables can be explained by a set of explanatory variables. This makes it particularly useful in soil science, where the interactions between soil properties, microbial communities, and plant nutrition are complex and interdependent (Do et al. 2024).

Imagine you are studying the chemical properties of soil, such as pH, organic matter content, and nutrient concentrations (like nitrogen, phosphorus, and potassium). At the same time, you might be interested in how these properties are influenced by environmental factors such as rainfall, temperature, or land use practices. Using RDA, you can analyze how much of the variation in soil chemical properties can be attributed to these environmental factors. For example, RDA might show that rainfall is a significant driver of soil organic matter content, while temperature primarily influences soil pH. This helps in understanding which factors are most important in shaping soil chemistry (Jia et al. 2024a).

In the context of soil physical properties, such as soil texture, bulk density, and moisture content, RDA can be used to link these properties with different land management practices. For instance, you might want to investigate how different types of tillage or irrigation practices impact soil structure. By applying RDA, you could determine which practices lead to more favorable physical conditions for crop growth. For example, RDA could reveal that conservation tillage practices are associated with higher soil moisture content and lower bulk density, which are critical for maintaining soil health and fertility (Jia et al. 2024b).

RDA is also powerful when studying soil microbial communities, which are essential for nutrient cycling, soil structure, and plant health. Microbial communities in the soil can be influenced by a variety of factors, including soil chemical properties, plant species, and environmental conditions. By using RDA, you can explore how variations in microbial community composition are related to different soil chemical properties. For example, RDA might show that certain microbial taxa are strongly associated with high nitrogen levels, while others are linked to acidic soils. This can help in identifying key factors that promote beneficial microbial communities in the soil (Do et al. 2024).

Plant nutrition is another area where RDA can provide valuable insights. When studying how different levels of nutrients, such as copper (Cu) and zinc (Zn), affect plant growth and nutrient uptake, RDA allows you to assess how these nutrients interact with each other and with other soil properties. For instance, RDA might reveal that Zn levels have a stronger impact on plant nutrient uptake in soils with high organic matter, while Cu levels are more influential in soils with lower pH. This kind of analysis can inform better fertilizer management practices, ensuring that plants receive the right balance of nutrients for optimal growth (Jia-nan et al. 2024).

One of the key concepts to understand with RDA is that it not only summarizes the variation within the response variables (such as soil properties or microbial communities) but also explains this variation in the context of the explanatory variables (such as environmental factors or management practices). This dual focus makes RDA a powerful tool for hypothesis testing, allowing researchers to determine which factors are driving changes in their systems of interest. In soil science, where multiple variables are often at play, this can be particularly valuable for making informed decisions about land management and conservation (Kaoping et al. 2024).

Moreover, RDA provides a clear and visual way to interpret complex data. The results of an RDA analysis are typically presented as ordination diagrams, where the axes represent the variation in the response variables that can be explained by the explanatory variables. These diagrams help in visualizing how different variables are related and which explanatory variables are most influential. For young scientists, learning to interpret these diagrams is crucial for understanding the complex interactions within soil ecosystems and applying this knowledge in practical settings (Li et al. 2024b).

In summary, RDA is an essential tool for soil scientists because it allows for the simultaneous analysis of multiple variables and provides insights into how different factors interact within soil ecosystems. Whether you are studying soil chemical properties, physical properties, microbial communities, or plant nutrition, RDA can help you identify the key drivers of change and inform more effective management practices. By mastering RDA, young scientists can contribute to the sustainable management of soils, ensuring their health and productivity for future generations.

Main Assumptions Required for RDA

Redundancy Analysis (RDA) is a powerful multivariate technique, but its effectiveness and reliability depend on meeting several key assumptions. These assumptions are crucial because they ensure that the relationships identified by RDA are meaningful and not artifacts of the data or statistical process. When these assumptions are met, RDA can provide valuable insights into the relationships between response and explanatory variables, particularly in fields like soil science (Liu et al. 2024).

One of the primary assumptions of RDA is linearity. RDA assumes that the relationships between the response variables (such as soil properties) and the explanatory variables (such as environmental factors or soil types) are linear. This means that changes in the explanatory variables lead to proportional changes in the response variables. For example, in soil science, if you are studying the impact of sand content on soil pH and nutrient availability, RDA assumes that an increase in sand content will result in a linear change in these soil properties. However, if the relationship is nonlinear (e.g., soil pH changes more drastically at certain levels of sand content), RDA might not accurately capture the true dynamics of the system (Kaoping et al. 2024).

Another important assumption is the independence of observations. RDA assumes that the observations in the dataset are independent of each other. This means that the response of one observation should not influence the response of another. In the context of soil science, this could be a challenge if samples are taken from locations that are spatially close to each other, as neighboring soil samples might be more similar due to shared environmental conditions. For instance, in sandy soils versus clay soils, the properties of neighboring sandy soil samples might be more alike due to similar drainage patterns or root penetration, potentially violating the assumption of independence (Liu et al. 2024; Wu et al. 2024).

RDA also assumes that the data are multivariate normal. This assumption is less strict than in some other statistical methods, but it still plays a role in the reliability of the analysis. Multivariate normality means that the response variables should follow a normal distribution when considered together as a group. In soil science, this might involve variables like soil organic matter, pH, and nutrient concentrations. If these variables are heavily skewed or have outliers, it could impact the results of the RDA. For example, in comparing sandy soils to clay soils, the distribution of nutrient concentrations might be different due to the distinct properties of these soil types, potentially complicating the assumption of multivariate normality (Xiao et al. 2024).

The assumption of homogeneity of variance is also important in RDA. This means that the variance of the response variables should be consistent across the levels of the explanatory variables. In soil science, this could mean that the variability in soil pH or organic matter content should be similar across different types of soil, whether sandy or clayey. However, in practice, sandy soils often have more variable nutrient retention and drainage compared to clay soils, which could lead to heteroscedasticity (unequal variance) in the data. If this assumption is violated, it can lead to biased estimates and affect the interpretation of the RDA results (Li et al. 2024b).

Another key assumption is that the explanatory variables are not collinear. Collinearity occurs when two or more explanatory variables are highly correlated with each other, which can distort the results of the RDA by inflating the variance explained by these variables. In soil science, this might occur if you are using both soil texture (percentage of sand, silt, and clay) and soil porosity as explanatory variables since these properties can be closely related. For instance, sandy soils typically have higher porosity compared to clay soils. If these variables are included in the same RDA model, it could lead to misleading conclusions about their individual effects on the response variables (Zhao et al. 2024b).

Finally, RDA assumes that the relationship between the response and explanatory variables is consistent across the study area or system being analyzed. This assumption is particularly relevant in soil science, where the effects of environmental factors on soil properties might vary across different soil types. For example, the relationship between rainfall and soil moisture content might be different in sandy soils compared to clay soils due to their contrasting water retention capacities. If this spatial or contextual variability is not accounted for, the RDA results might oversimplify the actual dynamics at play (Zhang et al. 2024).

By understanding and addressing these assumptions, researchers can ensure that their use of RDA yields accurate and meaningful results. In soil science, where the complexity of interactions between soil types, environmental factors, and management practices is high, carefully considering these assumptions is crucial for drawing reliable conclusions about the factors that influence soil health and productivity (Xiong et al. 2024).

Steps to Conduct RDA

Conducting Redundancy Analysis (RDA) involves several systematic steps, each crucial for obtaining meaningful and accurate results. For young scientists in soil science and plant production, understanding these steps can help in analyzing complex datasets where multiple soil properties and plant nutrients are influenced by various environmental and management factors. Let's walk through the process using an example of comparing soil physical properties and plant nutrients in organic soils versus degraded soils (Xie et al. 2024).

(a) Step 1—Define the Research Question and Hypotheses: The first step in conducting RDA is to clearly define your research question and the hypotheses you want to test. This involves identifying the response variables (the variables you are interested in explaining, such as soil moisture content, bulk density, or plant nutrient levels) and the explanatory variables (the factors you believe might influence the response variables, such as soil type, organic matter content, or management practices). For instance, you might want to investigate how soil physical properties and nutrient availability differ between organic soils and degraded soils and how these differences affect plant growth.

(b) Step 2—Prepare the Data: Once your research question is clear, the next step is data collection and preparation. This involves gathering data on the response variables and explanatory variables for each observation (e.g., different soil samples or plots). In our example, you would collect data on soil moisture, bulk density, and nutrient concentrations (such as nitrogen, phosphorus, and potassium) for both organic and degraded soils. It's important to ensure that the data are complete, properly formatted, and free from errors. Missing data should be addressed, and variables should be standardized if they are on different scales.

(c) Step 3—Check the Assumptions: Before performing RDA, it's essential to check that the assumptions required for RDA are met. These include assumptions about linearity, multivariate normality, homogeneity of variance, and the absence of multicollinearity among the explanatory variables. For example, you should check that the relationships between soil physical properties and explanatory variables like organic matter content or soil type are linear. You should also verify that the variance in soil nutrient levels is consistent across different soil types and that the explanatory variables are not too highly correlated.

(d) Step 4—Perform the RDA: With your data prepared and assumptions checked, you can now perform the RDA. This step involves using statistical software to input your data and run the RDA analysis. The software will generate an ordination plot that shows how much of the variation in your response variables is explained by the explanatory variables. In our example, RDA might reveal how much of the variation in soil moisture and nutrient availability can be attributed to differences between organic and degraded soils. The analysis will also produce scores that indicate the strength and direction of these relationships.

(e) Step 5—Interpret the Results: After running the RDA, it's time to interpret the results. The ordination plot generated by the RDA will show the relationships between your response and explanatory variables. In the context of soil science, you might see that organic soils are associated with higher moisture content and nutrient levels, while degraded soils might be linked to higher bulk density and lower nutrient availability. The direction and magnitude of the arrows in the ordination plot can help you understand which explanatory variables are most influential and how they relate to the response variables.

(f) Step 6—Assess the Significance: RDA also allows you to test the statistical significance of the relationships identified in the analysis. This involves running permutation tests to determine whether the observed relationships are stronger than would be expected by chance. For example, you can test whether the differences in soil physical properties and plant nutrient levels between organic and degraded soils are statistically significant. This step is crucial for validating your findings and ensuring that they are robust.

(g) Step 7—Report the Findings: Once the analysis is complete and the results have been interpreted, the next step is to report your findings. This involves writing up your results in a clear and concise manner, including the ordination plot, the significance tests, and a discussion of what the results mean in the context of your research question. In our example, you would discuss how the differences in soil properties between organic and degraded soils impact plant nutrient availability and what implications this might have for soil management practices in agricultural systems.
(h) Step 8—Make Practical Recommendations: Finally, based on your findings, you can make practical recommendations for soil management and plant production. If your RDA analysis reveals that organic soils provide better conditions for plant growth due to higher moisture and nutrient levels, you might recommend practices that enhance soil organic matter content in degraded soils. These recommendations could include strategies like adding organic amendments, reducing tillage, or implementing cover cropping to improve soil health and increase productivity.

Handling High-Dimensional Soil Data with RDA

Redundancy Analysis (RDA) is particularly well-suited for handling high-dimensional soil data, which often involves numerous variables that are interrelated and can influence one another. In soil science, when studying all essential macronutrients (such as nitrogen, phosphorus, and potassium) alongside key soil properties (like pH, organic matter content, and texture), RDA helps to manage and interpret the complexity of these data by reducing dimensionality while retaining the relationships between the variables (Xiong et al. 2024).

When dealing with high-dimensional data, one challenge is the presence of many response variables (e.g., multiple macronutrient concentrations) that need to be analyzed in relation to several explanatory variables (e.g., soil properties). RDA addresses this by finding linear combinations of the response variables that maximize the amount of variation explained by the explanatory variables. This process effectively reduces the number of dimensions by summarizing the most important patterns in the data while still linking these patterns to specific soil properties (Li et al. 2024a, b; Xu et al. 2024).

For example, imagine you are studying a dataset that includes the concentrations of nitrogen, phosphorus, and potassium in various soil samples, along with soil pH, organic matter, and texture. Without RDA, it might be challenging to identify how these macronutrients are jointly influenced by the soil properties, especially when the data are complex and multidimensional. RDA simplifies this by transforming the original nutrient data into a smaller number of synthetic variables (canonical axes) that capture the major gradients in nutrient variation explained by the soil properties (Kaoping et al. 2024).

One of the advantages of RDA in handling high-dimensional soil data is its ability to provide clear visualizations. After performing the RDA, you can create ordination plots that illustrate how macronutrient concentrations are related to soil properties. For instance, the ordination plot might show that soils with higher organic matter tend to have higher nitrogen and phosphorus levels, while soils with lower pH are associated with lower potassium availability. This visualization helps to make sense of the complex interactions between the nutrients and soil properties in a more digestible form (Xu et al. 2024).

Another way RDA handles high-dimensional data is through the identification of key explanatory variables that account for most of the variation in the response variables. In the context of soil science, this might mean that among several soil properties, only a few, such as pH and organic matter content, are found to be the most significant in explaining the variation in macronutrient concentrations. By focusing on these key variables, RDA allows researchers to simplify the analysis without losing critical information, making it easier to draw meaningful conclusions from complex datasets (Xue et al. 2024).

Finally, RDA is robust to situations where some of the variables might be highly correlated, a common issue in high-dimensional data. For example, in soils, organic matter and nutrient content often correlate, as organic matter can be a major source of nutrients. RDA manages these correlations by integrating them into the canonical axes, ensuring that the relationships are accurately reflected in the analysis. This reduces the risk of multicollinearity distorting the results, providing a clearer understanding of how soil properties and macronutrients interact. Through these mechanisms, RDA proves to be a powerful tool in soil science, enabling researchers to navigate and interpret high-dimensional datasets effectively. By focusing on the most important relationships and reducing complexity, RDA helps to uncover the key factors that influence soil health and fertility, even when the data are complex and multidimensional (Bi et al. 2024).

Why RDA Is Particularly Useful in Soil Science Research?

Redundancy Analysis (RDA) is particularly useful in soil science research because it allows for the analysis of complex, multivariate data where multiple soil properties and environmental factors interact. Soil systems are inherently complex, with numerous physical, chemical, and biological variables influencing each other. RDA provides a robust framework for understanding these interactions by linking variations in soil properties to specific environmental or management factors. This makes it especially valuable in studies where understanding the influence of multiple variables on soil health and function is critical (Chen et al. 2024b).

In the context of saline soils, RDA can be particularly insightful. Saline soils are characterized by high concentrations of soluble salts, which can adversely affect soil structure, nutrient availability, and plant growth. Studying saline soils often involves examining how various soil properties, such as electrical conductivity (EC), pH, sodium adsorption ratio (SAR), and nutrient levels (e.g., nitrogen, phosphorus, potassium), interact and are influenced by factors like irrigation practices, soil texture, and land management. RDA is ideal for analyzing these relationships because it can handle multiple response variables (like soil nutrient concentrations and EC) and explanatory variables (such as soil texture and irrigation methods) simultaneously (Jia et al. 2024a).

For example, a researcher might want to investigate how different irrigation practices affect soil salinity and nutrient availability in saline soils. By applying RDA, the researcher could analyze how variations in irrigation frequency and water quality (explanatory variables) are related to changes in soil EC, SAR, and nutrient levels (response variables). The RDA could reveal that certain irrigation practices are associated with higher EC and SAR, indicating an increase in soil salinity, while other practices might help to maintain nutrient availability despite the saline conditions. This information is critical for developing sustainable management strategies for saline soils (Li et al. 2024a).

RDA is also useful in identifying the most influential factors driving changes in saline soil properties. For instance, in a study of saline soils across different regions, RDA could help determine whether soil texture (e.g., sand vs. clay content) or irrigation practices have a more significant impact on soil salinity and nutrient dynamics. The ordination plot produced by the RDA would show the relationship between these factors, allowing researchers to focus on the most critical variables affecting soil health in saline environments. Furthermore, RDA's ability to visualize complex data makes it particularly beneficial in soil science. In the case of saline soils, the ordination plot generated by RDA can help researchers and land managers easily understand how different soil properties are related to salinity levels and how these relationships vary under different management practices. This visual representation can be a powerful tool for communicating findings to stakeholders, such as farmers or policymakers, who need to make informed decisions about land use and soil management in saline environments (Mu et al. 2024).

Lastly, RDA is valuable for hypothesis testing in soil science research. Researchers can use RDA to test specific hypotheses about the effects of salinity on soil properties and plant growth. For example, they might hypothesize that high salinity reduces the availability of essential nutrients like potassium and magnesium. By using RDA, they can quantify the extent to which salinity levels explain the variation in nutrient availability, providing empirical support for their hypothesis and guiding further research or management interventions. In summary, RDA is particularly useful in soil science research because it allows for the comprehensive analysis of multiple interacting variables, provides clear visualizations of complex data, and supports hypothesis testing. In the study of saline soils, RDA can help researchers understand the intricate relationships between soil properties and environmental factors, leading to more effective management strategies for maintaining soil health and productivity in challenging conditions (Peng et al. 2024).

Practical Application in Agroforestry vs. Natural Ecosystems

The practical application of Redundancy Analysis (RDA) in comparing agroforestry and conventional farming systems involves exploring how different soil management practices impact soil health, productivity, and sustainability. Agroforestry, which integrates trees with crops and livestock, often leads to more complex and diverse interactions in the soil compared to conventional farming, which typically focuses on monocultures and intensive use of inputs like fertilizers and pesticides. RDA can be used to assess these differences by analyzing how various soil properties respond to the different management practices in these systems (Wu et al. 2024).

In agroforestry systems, soil management practices often include minimal tillage, organic amendments, and the incorporation of tree litter, which can enhance soil organic matter, nutrient cycling, and microbial diversity. By applying RDA, researchers can investigate how these practices influence soil properties such as organic carbon content, pH, and nutrient availability. For instance, RDA might reveal that soils under agroforestry have higher organic matter and better nutrient retention compared to conventional farming systems, where soil disturbance and synthetic inputs can lead to nutrient losses and soil degradation (Xiong et al. 2024).

RDA can also help in understanding the impacts of tree species diversity in agroforestry systems on soil health. Different tree species contribute varying amounts of organic matter and root exudates to the soil, which can affect soil structure, microbial communities, and nutrient dynamics. By using RDA, researchers can assess how the diversity of tree species in an agroforestry system correlates with improvements in soil properties such as increased soil moisture retention or enhanced nutrient availability. This analysis can help identify which tree species or combinations thereof are most beneficial for maintaining soil health (Jia et al. 2024a).

In conventional farming systems, soil management practices often involve the use of synthetic fertilizers, pesticides, and frequent tillage, which can have different effects on soil properties compared to agroforestry. RDA can be applied to analyze how these practices influence soil characteristics like nutrient content, soil compaction, and microbial activity. For example, RDA might show that while conventional farming systems with high fertilizer inputs have higher immediate nutrient levels, they may also suffer from reduced microbial diversity and increased soil compaction, leading to long-term soil degradation (Chang et al. 2024).

Another practical application of RDA in comparing these systems is in assessing the long-term sustainability of soil management practices. Agroforestry systems, with their focus on maintaining soil cover and organic inputs, often promote more sustainable soil health over time. RDA can help quantify these benefits by linking the stability of soil properties like organic matter content and pH to long-term management practices in both agroforestry and conventional systems. For instance, RDA might demonstrate that soils under agroforestry are more resilient to erosion and nutrient depletion compared to those under conventional farming, where the soil is more vulnerable to degradation (Bi et al. 2024).

RDA also allows for the evaluation of soil biological properties, such as microbial biomass and enzyme activities, which are critical indicators of soil health. In agroforestry systems, the diversity of plant species and organic inputs can enhance microbial activity and diversity, leading to improved soil fertility. By applying RDA, researchers can explore how these biological properties vary between agroforestry and conventional farming systems and how they are influenced by different management practices. For example, RDA could show that agroforestry practices lead to higher microbial biomass and enzyme activity, contributing to more efficient nutrient cycling and soil health (Jia-nan et al. 2024).

The visual output of RDA, such as ordination plots, provides a clear representation of how soil properties respond to different management practices in agroforestry versus conventional farming systems. These visualizations can be particularly useful for communicating the benefits of agroforestry to farmers, policymakers, and other stakeholders. By showing how agroforestry practices lead to improved soil health compared to conventional methods, RDA can help advocate for more sustainable farming practices that protect and enhance soil resources. The practical application of RDA in comparing agroforestry and conventional farming systems lies in its ability to analyze and visualize the complex interactions between soil management practices and soil health. By using RDA, researchers can demonstrate the long-term benefits of agroforestry practices, such as improved soil structure, nutrient retention, and biological activity, compared to the more immediate but potentially unsustainable impacts of conventional farming. This analysis provides valuable insights that can guide the adoption of more sustainable soil management practices in agricultural systems (Li et al. 2024b).

Applying RDA in Soil Science

Redundancy Analysis (RDA) is a valuable tool for young scientists in soil science, food production, and plant nutrition, particularly when dealing with complex datasets involving multiple soil properties and management practices. RDA allows researchers to examine how different soil characteristics, such as physical and chemical properties, are influenced by various factors, making it easier to understand the intricate relationships that govern soil health and productivity. Let's explore how RDA can be applied in soil science using examples related to soil physical properties, chemical properties, and soil organic carbon pools in sandy soils under different management practices (Li et al. 2024a).

In sandy soils, physical properties like bulk density, soil texture, and porosity play a crucial role in determining water retention, root penetration, and nutrient availability. RDA can be used to analyze how these physical properties respond to different soil management practices, such as no-till farming, cover cropping, or the application of organic amendments. For example, RDA might reveal that no-till farming practices, which minimize soil disturbance, are associated with lower bulk density and improved porosity in sandy soils, leading to better water infiltration and root growth.

Soil chemical properties, including pH, cation exchange capacity (CEC), and nutrient concentrations, are also critical for maintaining soil fertility and supporting plant growth. RDA can help identify how these chemical properties are influenced by management practices in sandy soils. For instance, an RDA analysis might show that the application of organic amendments, such as compost or manure, increases soil CEC and nutrient availability, while also buffering soil pH in sandy soils. This information is valuable for designing management strategies that enhance soil fertility and crop production.

One of the key challenges in managing sandy soils is maintaining soil organic carbon (SOC) levels, which are essential for soil structure, nutrient cycling, and overall soil health. RDA can be applied to investigate how different management practices affect the SOC pools in sandy soils. For example, RDA might demonstrate that cover cropping, which involves planting crops like legumes or grasses during the off-season, significantly increases the SOC levels by adding organic matter to the soil. This can lead to improvements in soil structure and long-term fertility, particularly in sandy soils that are prone to organic matter depletion.

RDA is particularly useful for identifying which management practices are most effective at improving specific soil properties. In sandy soils, where issues like low water retention and nutrient leaching are common, RDA can help determine which practices, such as the use of organic mulches or reduced tillage, are most beneficial. For example, RDA might reveal that reduced tillage combined with the application of organic mulches is associated with higher soil moisture retention and reduced nutrient leaching, leading to better plant growth and higher yields.

Another advantage of RDA in soil science is its ability to handle high-dimensional data, where multiple soil properties and management practices are considered simultaneously. This is particularly important in sandy soils, where interactions between physical, chemical, and biological properties can be complex. By applying RDA, researchers can reduce the dimensionality of the data, focusing on the most significant relationships between soil properties and management practices. This simplification helps in drawing clearer conclusions and making more informed decisions about soil management.

RDA also allows for the visualization of these complex relationships through ordination plots. These plots can show how different soil properties are related to each other and to various management practices, making it easier to communicate findings to stakeholders. For instance, an ordination plot might clearly illustrate how the combination of cover cropping and organic amendment application leads to improved soil structure, higher nutrient levels, and increased SOC in sandy soils. This visual representation can be a powerful tool for educating farmers, policymakers, and other stakeholders about the benefits of sustainable soil management practices.

In addition to improving soil properties, RDA can also be used to assess the impact of soil management practices on plant nutrition and food production. For example, RDA might show that certain practices, such as the application of biochar or compost, not only enhance soil fertility in sandy soils but also lead to higher concentrations of essential nutrients in crops. This is crucial for ensuring that food production systems are both sustainable and capable of meeting the nutritional needs of the population.

Furthermore, RDA can help in understanding the long-term effects of soil management practices on soil health. By analyzing how soil properties change over time under different management regimes, researchers can identify practices that promote soil sustainability. For example, RDA might reveal that continuous application of organic amendments over several years leads to a gradual increase in SOC levels and improvements in soil structure, which are essential for maintaining soil productivity in sandy soils.

RDA is also valuable for hypothesis testing in soil science research. Researchers can use RDA to test specific hypotheses about the effects of management practices on soil properties. For instance, they might hypothesize that reduced tillage combined with organic amendments increases SOC levels in sandy soils more effectively than conventional tillage. By applying RDA, they can quantify the extent to which these practices explain the variation in SOC levels, providing empirical support for their hypothesis and guiding further research or management interventions. RDA is a powerful tool for young scientists in soil science, food production, and plant nutrition. By applying RDA, researchers can gain valuable insights into how different soil management practices affect soil physical and chemical properties, SOC pools, and overall soil health in sandy soils. This knowledge is essential for developing sustainable soil management strategies that enhance productivity, support plant nutrition, and ensure the long-term viability of food production systems.

Software for RDA

Several tools are commonly used for performing Redundancy Analysis (RDA). These tools vary in features, ease of use, and the level of statistical support they offer. Here's an overview of some popular options:

1. R (RStudio): R is a powerful and flexible open-source statistical software environment that includes extensive packages for multivariate analysis, including RDA. The vegan package in R is particularly well-suited for ecological and environmental data analysis, including RDA.

Key Features:

Flexibility: Allows for customization and complex analyses.

Extensive Packages: The vegan package includes functions for RDA, along with tools for data manipulation, visualization, and statistical testing.

Visualization: Capable of creating high-quality plots and ordination diagrams.

Example Use: To perform RDA in R, you might use commands such as rda() from the vegan package, followed by plot() to visualize the results.

2. Canoco: Canoco is a dedicated software package for multivariate analysis of ecological data, including RDA. It is known for its user-friendly interface and strong focus on ecological research.

Key Features:

User Interface: Provides a graphical interface that simplifies the process of performing and interpreting RDA.

Visualization Tools: Includes tools for creating various ordination plots and biplots.

Statistical Analysis: Offers extensive options for permutation tests and other statistical methods relevant to ecological data.

Example Use: Users can load their data into Canoco, select RDA from the analysis menu, and use the graphical interface to explore and visualize the results.

3. PRIMER: It is a software suite for analyzing ecological and environmental data, with RDA being one of the many available analyses. It is well-regarded for its ease of use and powerful visualization capabilities.

Key Features:

User-Friendly: Designed for users who may not have extensive statistical backgrounds.

Comprehensive Analysis: Includes tools for a wide range of multivariate analyses, not just RDA.

Graphical Output: Provides high-quality visualizations and plots.

Example Use: In PRIMER, users can import their data, choose RDA from the analysis menu, and use the provided tools to interpret and visualize the results.

4. PAST: PAST (Paleontological Statistics) is a free software package that provides a range of statistical tools, including RDA. It is commonly used in various scientific fields, including ecology and geology.

Key Features:

Accessibility: Free and open-source, making it accessible to a wide audience.

Diverse Functions: Offers a range of statistical analyses and graphical tools.

Ease of Use: Relatively straightforward interface suitable for beginners.

Example Use: Users can input their data, select RDA from the available analyses, and generate ordination plots and biplots to interpret the relationships between variables.

5. MVSP: MVSP (Multivariate Statistical Package) is another specialized software for multivariate analysis, including RDA. It is designed for users in various scientific fields requiring multivariate data analysis.

Key Features:

Specialized Tools: Provides specific tools for ordination and clustering, including RDA.

Visualization: Includes options for creating detailed plots and diagrams.

Support: Offers extensive documentation and user support.

Example Use: In MVSP, users can prepare their data, run RDA, and use the built-in tools to visualize and interpret the results.

6. SYSTAT: It is a comprehensive statistical software package that supports a range of statistical analyses, including RDA. It is known for its robust analytical capabilities and user-friendly interface.

Key Features:
Versatile Analysis: Supports a wide range of statistical methods beyond RDA.
Graphical Interface: Provides tools for creating various plots and diagrams.
Advanced Features: Offers advanced options for data manipulation and statistical testing.
Example Use: Users can import their datasets into SYSTAT, choose RDA from the analysis menu, and use the graphical interface to generate and interpret the results.

7. STATISTICA: It is a software suite that offers a variety of statistical and data analysis tools, including RDA. It is known for its comprehensive analytical capabilities and visualization options.

Key Features:
Broad Capabilities: Includes a wide range of statistical analyses and data manipulation tools.
User-Friendly: Provides an intuitive interface for performing analyses and generating plots.
Visualization: Capable of producing high-quality visual outputs.
Example Use: In STATISTICA, users can perform RDA by selecting the appropriate analysis options, and then utilize the software's tools for interpreting and visualizing the results.
Each of these software options has its strengths and is suited to different needs and levels of expertise. R is highly flexible and powerful but requires familiarity with coding. Canoco and PRIMER are user-friendly with strong visualization capabilities. PAST, MVSP, SYSTAT, and STATISTICA offer various levels of functionality and ease of use. Choosing the right tool depends on your specific needs, preferences, and familiarity with the software.

Example Application in Soil Science

Redundancy Analysis (RDA) can be a powerful tool for soil scientists looking to understand the relationships between soil properties and various influencing factors. Consider a scenario where a researcher is studying the impact of different land management practices on soil health in agricultural fields. By applying RDA, the researcher can analyze how soil physical properties, chemical properties, and organic carbon pools are influenced by these practices, providing valuable insights into optimal soil management strategies.

In this study, the researcher collects data from several fields with different management practices, such as conventional tillage, no-till farming, and organic farming. The dataset includes measurements of soil physical properties like bulk density, soil texture, and porosity, as well as chemical properties such as pH, cation exchange capacity (CEC), and nutrient concentrations. Soil organic carbon (SOC) levels are also recorded, as they are critical for assessing soil health and fertility.

The researcher uses RDA to examine how these soil properties respond to the different management practices. RDA helps to visualize the relationships between the response variables (soil properties) and explanatory variables (management practices). For instance, RDA might reveal that no-till farming is associated with lower bulk density and higher soil porosity compared to conventional tillage, which can lead to better water infiltration and root growth.

In terms of soil chemical properties, RDA can show how different management practices affect soil pH and nutrient availability. For example, the analysis might indicate that organic farming practices lead to higher levels of soil organic matter and improved CEC, resulting in better nutrient retention and availability. This information is essential for understanding how different practices impact soil fertility and plant growth. One of the significant benefits of RDA in this context is its ability to handle multiple response variables simultaneously. For example, the researcher might be interested in how soil organic carbon pools are influenced by different management practices. RDA can help identify which practices lead to higher SOC levels and how this correlates with other soil properties such as pH and nutrient concentrations.

The RDA results can be visualized using ordination plots, which provide a clear representation of how soil properties are related to management practices. For instance, a biplot might show that fields under organic farming practices cluster together, indicating similar soil properties, while fields under conventional tillage display a different pattern. This visualization helps to communicate the findings effectively to stakeholders and guide decision-making.

Additionally, RDA can be used to assess the impact of specific variables on soil health. For example, the researcher might use RDA to evaluate how soil texture (sand, silt, clay content) affects the relationship between management practices and soil organic carbon pools. The analysis could reveal that soils with higher clay content respond differently to management practices compared to sandy soils, providing insights into how soil texture influences the effectiveness of different practices.

In a broader context, RDA can help in comparing different regions or land types. For instance, the researcher might analyze soil data from fields located in different climatic zones to see how management practices affect soil properties across these zones. RDA can show whether certain practices are more effective in specific regions and help tailor management strategies to local conditions. RDA also supports hypothesis testing in soil science research. For example, the researcher might hypothesize that no-till farming leads to better soil health compared to conventional tillage due to its impact on soil structure and organic matter. By applying RDA, the researcher can quantify the extent to which no-till farming explains variations in soil properties and test the validity of this hypothesis.

Lastly, RDA provides a framework for developing and implementing sustainable soil management practices. By understanding the relationships between management practices and soil properties, researchers can recommend practices that improve soil health and productivity. For example, the findings might suggest that incorporating cover crops and reducing tillage are effective strategies for enhancing soil organic carbon and overall soil quality. It is a versatile tool in soil science that helps researchers analyze and visualize the complex relationships between soil properties and management practices. By applying RDA, scientists can gain valuable insights into how different practices affect soil health and productivity, ultimately guiding the development of more sustainable soil management strategies.

Conclusion

Redundancy Analysis (RDA) stands as a vital tool in soil science and data analysis due to its ability to elucidate complex relationships between multiple soil properties and influencing factors. By effectively reducing the dimensionality of high-dimensional datasets, RDA enables researchers to discern how various soil management practices, environmental conditions, and physical and chemical properties interact and influence soil health. Its strength lies in its capacity to handle both response and explanatory variables simultaneously, providing clear visualizations and statistical insights that facilitate a deeper understanding of soil dynamics. This capability is crucial for developing informed, sustainable soil management strategies that enhance soil fertility, optimize nutrient availability, and support agricultural productivity. Ultimately, RDA supports robust decision-making and policy development by translating complex soil data into actionable knowledge, driving improvements in soil management practices and contributing to the advancement of soil science.

References

Bi Y, Guo N, Zhang Y et al (2024) Responses of soil stoichiometry and soil enzyme activities in the different distance around opencast coal mine of the Hulun Buir Grassland of China. Int J Coal Sci Technol 11:14. https://doi.org/10.1007/s40789-024-00662-4

Chang D, Zhang H, Zhu Z et al (2024) Effects of soil conditioners on vineyard saline soil physicochemical properties and bacterial community in arid areas. Land Degrad Dev 35:3434–3447. https://doi.org/10.1002/ldr.5143

Chen CN, Liao CS, Tzou YM et al (2024a) Soil quality and microbial communities in subtropical slope lands under different agricultural management practices. Front Microbiol 14:1242217. https://doi.org/10.3389/fmicb.2023.1242217

Chen Y, Jiang Z, Ou J et al (2024b) Nitrogen substitution practice improves soil quality of red soil (Ultisols) in South China by affecting soil properties and microbial community composition. Soil Tillage Res 240:106089. https://doi.org/10.1016/j.still.2024.106089

Do MTT, Van LN, Le XH et al (2024) National variability in soil organic carbon stock predictions: impact of bulk density pedotransfer functions. Int Soil Water Conserv Res. https://doi.org/10.1016/j.iswcr.2024.04.002

Jia W, Huang P, Zhu K et al (2024a) Zonation of bulk and rhizosphere soil bacterial communities and their covariation patterns along the elevation gradient in riparian zones of three Gorges reservoir, China. Environ Res 249:118383. https://doi.org/10.1016/j.envres.2024.118383

Jia Z, Li C, Ma S et al (2024b) Mineral-solubilizing microbial inoculums promote robinia Pseudoacacia L. growth by optimizing rhizospheric soil microbial community structure. J Soil Sci Plant Nutr. https://doi.org/10.1007/s42729-024-01965-w

Jia-nan W, Hao-ming F, Yan-feng J (2024) Spatial variation characteristics and influencing factors of sediment connectivity in the black soil region of northeast China. Geoderma 446:116895. https://doi.org/10.1016/j.geoderma.2024.116895

Kaoping Z, Song Y, Xu H et al (2024) Soil C:N:P stoichiometry regulates the response of soil bacterial community composition and putative functions to coastal reclamation. J Soils Sediments 24:819–828. https://doi.org/10.1007/s11368-023-03657-x

Li H, Dai X, Gao Z (2024a) Dependency of long-term soil quality controls on summer fallow tillage and soil layers for dryland winter wheat in loess plateau. Agriculture 14:1026. https://doi.org/10.3390/agriculture14071026

Li X, Gong Q, Li Z (2024b) Response characteristics of soil microorganisms under strong disturbance conditions in the riparian zone of the three Gorges reservoir Area. Sci Rep 14:18394. https://doi.org/10.1038/s41598-024-69533-x

Liu P, Ding S, Liu N et al (2024) Effects of detritus treatments on soil microbial community composition, structure and nutrient limitation in a subtropical karst ecosystem. J Soil Sci Plant Nutr 24:3265–3281. https://doi.org/10.1007/s42729-024-01750-9

Mu Y, Ye R, Xiong K et al (2024) Response of soil organic carbon to land-use change after farmland abandonment in the karst desertification control. Plant Soil 501:595–610. https://doi.org/10.1007/s11104-024-06541-w

Peng J, Wang J, Yang Q et al (2024) Spatial variation in soil aggregate stability and erodibility at different slope positions in four hilly regions of northeast China. Catena 235:107660. https://doi.org/10.1016/j.catena.2023.107660

Shao L, Yan P, Ye S et al (2024) Soil organic matter and water content affect the community characteristics of arbuscular mycorrhizal fungi in Helan mountain, an arid desert grassland area in China. Front Microbiol 15:1377763. https://doi.org/10.3389/fmicb.2024.1377763

Wu X, Yang Y, He T et al (2024) Temporal variability of saturated hydraulic conductivity on a typical black soil slope of northeast China. Catena 236:107742. https://doi.org/10.1016/j.catena.2023.107742

Xiao M, Chen C, Yao R, Wang X, Liu G (2024) Response of soil fungal community in coastal saline soil to short-term water management combined with bio-organic fertilizer. Agronomy 14:1441. https://doi.org/10.3390/agronomy14071441

Xie J, Huang R, Wu D et al (2024) Soil organic matter and amorphous oxides determine the surface electrochemical properties of purple soil under long-term inorganic and organic fertilization. J Soil Sci Plant Nutr. https://doi.org/10.1007/s42729-024-01941-4

Xiong R, He X, Gao N et al (2024) Soil pH amendment alters the abundance, diversity, and composition of microbial communities in two contrasting agricultural soils. Microbiol Spectr. https://doi.org/10.1128/spectrum.04165-23

Xu X, Wang J, Niu Y et al (2024) 44-years of fertilization altered soil microbial community structure by changing soil physical, chemical properties and enzyme activity. J Soil Sci Plant Nutr 24:3150–3161. https://doi.org/10.1007/s42729-024-01740-x

Xue D, Wang Y, Sun H et al (2024) Effects of soil conditioner (volcanic ash) on yield quality and rhizosphere soil characteristics of melon. Plants 13:1787. https://doi.org/10.3390/plants13131787

Zhang Y, Zhang R, Zhao S et al (2024) Impact of nitrogen use efficiency towards ammonia-oxidizing microbes in rhizosphere soil of intercropped soybean and maize. J Soil Sci Plant Nutr. https://doi.org/10.1007/s42729-024-01964-x

Zhao M, Sun Y, Liu S et al (2024a) Effects of stand density on the structure of soil microbial functional groups in Robinia pseudoacacia plantations in the hilly and gully region of the Loess Plateau, China. Sci Total Environ 912:169337. https://doi.org/10.1016/j.scitotenv.2023.169337

Zhao Y, Han Z, Zhang G et al (2024b) Variability of soil enzyme activities and nutrients with forest gap renewal interacting with soil depths in degraded karst forests. Ecol Indic 166:112332. https://doi.org/10.1016/j.ecolind.2024.112332

Zhu P, Chen S, Wang C, Xu Z (2024) Short-term conservation tillage degrades soil infiltration properties in the black soil region of Northeast China. Eur J Soil Sci 75:e13479. https://doi.org/10.1111/ejss.13479

Non-metric Multidimensional Scaling (NMDS)

6

Abstract

Non-metric Multidimensional Scaling (NMDS) is an ordination technique used to visualize complex, high-dimensional data by representing it in a lower-dimensional space while preserving the rank order of dissimilarities between samples. Unlike Principal Component Analysis (PCA), which assumes linear relationships and normality in the data, NMDS is non-parametric and handles non-linear relationships and non-normal distributions. The main assumptions for NMDS include using accurate dissimilarity or distance measures and ensuring that data is ordinal, as NMDS does not rely on specific distributional assumptions. NMDS effectively manages high-dimensional soil data by reducing it to a comprehensible form, making it easier to interpret complex interactions between soil properties, microbial communities, and environmental factors. Its utility in soil science stems from its ability to reveal patterns and relationships that are not easily captured by other methods, thereby informing better soil management and conservation strategies. Before applying NMDS, researchers should verify the appropriateness of their dissimilarity measures, ensure sufficient sample size to represent data variability, and check that their data is suitable for NMDS to obtain reliable and meaningful results.

Keywords

Complex data · High-dimensional data visualization · Non-metric multidimensional scaling · Ordination · Soil Science

Questions Covered in the Chapter

(a) What is the primary difference between NMDS and PCA?
(b) What are the main assumptions required for NMDS?

T. Souza, *Advanced Statistical Analysis for Soil Scientists*,
https://doi.org/10.1007/978-3-031-88161-9_6

(c) How does NMDS handle high-dimensional soil data?
(d) Why is NMDS particularly useful in soil science research?
(e) What should researchers check for before applying NMDS to ensure reliable results?

Introduction

Non-metric multidimensional scaling (NMDS) is a robust statistical technique used to visualize the similarity or dissimilarity of data points in a low-dimensional space. Unlike traditional scaling methods, which rely on preserving the exact distances between points, NMDS prioritizes the ranking of these distances, making it particularly useful for handling complex, non-linear data. In this technique, the goal is to represent the structure of data based on the ranks of pairwise dissimilarities, often calculated using ecological or environmental datasets. This flexibility allows NMDS to manage data that do not meet the assumptions required by other ordination methods, making it an attractive option for analyzing complex datasets, particularly those involving ecological or biological data (Botha et al. 2024).

One of the key features of NMDS is its minimal reliance on assumptions about the data's distribution. Unlike methods such as Principal Component Analysis (PCA) or Linear Discriminant Analysis (LDA), which assume linear relationships and normally distributed variables, NMDS only assumes that the data can be ordered by rank. This allows NMDS to be applied to datasets where the relationships among variables are unknown or where the data are non-linear. This assumption of rank order rather than distance makes NMDS particularly useful in fields where data are often non-linear or noisy, such as ecology, environmental science, and soil science (Huusko et al. 2024).

The advantages of NMDS are numerous, particularly when dealing with ecological or environmental data. One significant advantage is its ability to handle complex, multidimensional data and reduce it to a simpler form that is easier to interpret. This is especially useful in soil science, where researchers often deal with data from various sources, including chemical, physical, and biological measurements. NMDS allows for the integration of these diverse data types into a coherent analysis, making it easier to identify patterns and relationships that might be obscured in a more complex dataset. Another advantage of NMDS is its flexibility in handling different types of data. Whether the data are continuous, ordinal, or categorical, NMDS can be applied effectively, allowing for a wide range of analyses. This versatility is crucial in soil science, where data often come in various forms, from continuous measurements of soil properties to categorical data on land use types. NMDS allows researchers to analyze these disparate data types together, providing a more holistic view of the soil ecosystem (Lu et al. 2024).

NMDS is also valuable in advanced statistical analysis because of its ability to provide a visual representation of the data. The resulting ordination plot from NMDS allows researchers to see the relationships between data points clearly, facilitating the interpretation of complex datasets. This visual aspect of NMDS is particularly important in soil science, where understanding the relationships between different soil properties and environmental factors is crucial for making informed decisions about land management and conservation (Liu et al. 2024a).

In the context of soil science, NMDS plays a critical role in understanding the complex interactions between soil properties and environmental factors. Soil is a highly heterogeneous medium, with properties that can vary significantly over small spatial scales. NMDS allows researchers to analyze and visualize this complexity, helping to identify patterns and relationships that are not immediately apparent. This understanding is crucial for managing soil resources effectively, particularly in the face of challenges such as climate change and land degradation. The importance of NMDS in soil science is further underscored by its ability to handle large datasets with multiple variables. In modern soil science, researchers often collect vast amounts of data from various sources, including remote sensing, field measurements, and laboratory analyses. NMDS provides a powerful tool for reducing the complexity of these datasets and extracting meaningful insights. By identifying key patterns and relationships, NMDS can help guide soil management practices, ensuring that they are based on a solid understanding of the underlying soil processes (Ni et al. 2024).

Finally, NMDS is not just a tool for analysis but also a bridge between statistical methods and practical applications in soil science. By providing a clear, visual representation of complex data, NMDS helps translate abstract statistical results into actionable insights. This makes it an invaluable tool for soil scientists, agronomists, and environmental managers who need to make decisions based on complex and often conflicting data. Through its ability to simplify and clarify, NMDS contributes to a deeper understanding of soil ecosystems and supports the sustainable management of soil resources.

Main Differences Between NMDS and PCA

The primary difference between Non-metric Multidimensional Scaling (NMDS) and Principal Component Analysis (PCA) lies in how they approach the ordination of data and the assumptions they make about that data. PCA is a linear technique that assumes that the data has a multivariate normal distribution and seeks to reduce the dimensionality of the data by transforming it into a set of orthogonal (uncorrelated) axes that maximize the variance. NMDS, on the other hand, is a non-linear technique that does not make assumptions about the distribution of the data and instead focuses on preserving the rank order of distances between points in a lower-dimensional space. This fundamental difference in approach makes each method suitable for different types of data and different analytical goals (Souza et al. 2024).

PCA operates by projecting data onto a set of principal components that are linear combinations of the original variables. These principal components are chosen such that the first component accounts for the largest possible variance in the data, the second component accounts for the next largest variance, and so on. This linear transformation allows PCA to simplify the dataset while retaining as much of the variance as possible. However, this approach assumes that the relationships among the variables are linear, which can be a limitation when dealing with complex, non-linear data. For example, in the field of plant nutrition, if a researcher is studying the effects of varying levels of phosphorus (P) and zinc (Zn) on plant growth, PCA might oversimplify the relationships if the effects of P and Zn on plant growth are not linearly related (Wang et al. 2024a).

In contrast, NMDS does not rely on linear relationships or assumptions about the distribution of the data. Instead, NMDS focuses on preserving the rank order of pairwise dissimilarities between data points when reducing the data to a lower-dimensional space. This allows NMDS to handle non-linear and complex relationships more effectively than PCA. In the context of plant nutrition, if the relationship between P and Zn levels and plant growth is non-linear—perhaps because high levels of Zn interfere with P uptake in a non-linear manner—NMDS would be better suited to capture and visualize these complex interactions. NMDS would represent the data in a way that reflects the actual dissimilarities between the effects of different P and Zn levels on plant growth, even if these effects are non-linear (Xu et al. 2024).

The type of output produced by PCA and NMDS also differs significantly. PCA generates principal components that can be interpreted as axes along which the data varies most. These components are orthogonal, meaning that they are uncorrelated with each other. This makes PCA results relatively straightforward to interpret, particularly when the first few principal components explain a large proportion of the variance in the data. In the case of plant nutrition, a PCA might reveal that the first principal component is strongly associated with P levels, while the second component is associated with Zn levels. However, this interpretation hinges on the assumption of linearity, which may not fully capture the true dynamics of how P and Zn interact in influencing plant growth (Luo et al. 2024).

NMDS, on the other hand, produces an ordination plot where the axes are not based on linear combinations of the original variables but instead are derived from the configuration of points that best preserves the rank order of dissimilarities. The axes in an NMDS plot do not have a straightforward interpretation as in PCA. Instead, the focus is on the relative positions of the data points in the plot, which reflect the underlying structure of the data as determined by the dissimilarities. For instance, in plant nutrition research, an NMDS plot might reveal clusters of data points that correspond to specific combinations of P and Zn levels that produce similar effects on plant growth. This can be particularly useful when the interactions between P and Zn are complex and not easily captured by linear methods (Maciel et al. 2024).

Another critical difference lies in the sensitivity to data transformation. PCA is sensitive to the scaling of the data; therefore, it often requires that the data be standardized or normalized before analysis. This is because PCA maximizes the variance, and unscaled data could lead to components being dominated by variables with the largest scales. In the plant nutrition example, if the P and Zn concentrations are measured on vastly different scales, PCA might be biased toward the variable with the larger scale, potentially misrepresenting the true interaction between these nutrients. NMDS, however, operates on ranked dissimilarities, which makes it less sensitive to the original scaling of the data. This allows NMDS to more accurately reflect the true structure of the data without the need for extensive preprocessing (Mondal et al. 2024).

When considering data interpretation, PCA often provides a more direct and interpretable result, particularly when the first two or three components explain a significant portion of the variance. This can be advantageous in scenarios where a clear, linear relationship exists between variables, as it simplifies the task of understanding which variables are driving the observed patterns. In plant nutrition studies, this might mean quickly identifying that P levels are the primary driver of plant growth variation across treatments, with Zn playing a secondary role. However, this clarity comes at the cost of potentially missing out on more complex, non-linear relationships (Liu et al. 2024b).

In contrast, NMDS emphasizes the preservation of the relative dissimilarities between data points rather than the direct interpretation of variables. This can make NMDS more challenging to interpret, as the axes do not correspond to specific variables or combinations of variables. Instead, researchers must rely on the overall configuration of points and additional tools, such as vector fitting, to interpret the ordination plot. However, this flexibility allows NMDS to capture and represent complex, non-linear relationships more effectively. In plant nutrition, where nutrient interactions are often non-linear, NMDS might reveal subtle patterns that would be obscured by PCA, such as a non-linear threshold effect where Zn levels above a certain point dramatically reduce the availability of P to plants (Gómez et al. 2024).

Moreover, NMDS is often more robust in handling missing data or data with outliers, as it relies on rank order rather than exact distances. In ecological and agricultural studies, missing data can be common due to the complexity of field experiments. In such cases, NMDS can still provide meaningful insights without requiring imputation or other adjustments that might introduce bias. For example, if some P and Zn treatment combinations in a plant nutrition study have missing data points due to experimental issues, NMDS can still provide a reliable ordination that reflects the overall structure of the data, whereas PCA might be more sensitive to these gaps (Dunleavy and Mack 2024).

Ultimately, the choice between NMDS and PCA should be guided by the nature of the data and the specific research questions at hand. For linear, normally distributed data where the goal is to reduce dimensionality while retaining as much variance as possible, PCA is likely the better choice. However, for complex, non-linear data where preserving the rank order of dissimilarities is more important, NMDS offers a more flexible and powerful tool. In the context of plant nutrition, where interactions between nutrients like P and Zn can be complex and non-linear, NMDS may provide a more nuanced and accurate representation of the data, helping researchers to uncover subtle patterns and relationships that might otherwise be missed (Idbella et al. 2024).

Key Concepts of NMDS

Non-metric multidimensional scaling (NMDS) is a powerful tool used in soil science to explore and visualize complex ecological data, such as the composition of soil fungi communities. Soil ecosystems are incredibly diverse, with many species of fungi interacting with each other and with the environment. Understanding these interactions and patterns can be challenging, especially when dealing with large datasets that contain multiple variables. NMDS helps simplify this complexity by creating a visual representation of the data, making it easier to see patterns and relationships that might otherwise be hidden (Chen et al. 2024).

One of the main strengths of NMDS is its flexibility in handling data that do not follow a normal distribution or that involve non-linear relationships. Soil fungi communities are often influenced by a variety of factors, such as soil pH, moisture, organic matter content, and nutrient availability. These factors can interact in complex ways that are not always linear or straightforward. For example, certain fungi may thrive in acidic soils with high organic matter, while others may prefer neutral pH and lower organic content. NMDS allows scientists to visualize these complex relationships by preserving the rank order of differences between samples rather than relying on linear relationships, which might oversimplify the data (Chepkemoi et al. 2024).

In practical terms, NMDS generates an ordination plot, a type of scatter plot that represents the samples based on their dissimilarities in fungal community composition. Imagine you have collected soil samples from different parts of a forest and measured the types and abundance of fungi in each sample. You could use NMDS to create a plot where each point represents a soil sample, and the distance between points reflects how similar or different the fungal communities are in those samples. Samples with similar fungal communities will appear close together, while those with very different communities will be farther apart. This visual approach makes it easier to identify clusters of similar samples and potential environmental gradients influencing the fungal communities (Deng et al. 2024).

One example of how NMDS can be used in soil science is in studying the effects of soil pH on fungal community composition. Suppose you are investigating how fungi in the soil respond to different pH levels across a gradient from acidic to alkaline conditions. By applying NMDS, you could create an ordination plot showing the distribution of fungal communities along this pH gradient. You might observe that samples with similar pH levels cluster together, indicating that pH is a strong driver of fungal community structure. This insight can help you understand the ecological roles of different fungi and how they adapt to varying soil conditions (Islam et al. 2024).

NMDS is also beneficial when comparing the fungal communities from different land-use types, such as agricultural fields, forests, and grasslands. Each land-use type is likely to support a distinct set of fungi due to differences in soil management, vegetation, and microclimate. By applying NMDS to the fungal data, you can create a plot that visually separates the samples according to their land-use type. For instance, samples from forests might cluster together, reflecting a unique fungal community adapted to undisturbed, organic-rich soils, while samples from agricultural fields might form a different cluster due to the impact of tillage and fertilizers. This type of analysis is crucial for understanding how land use affects soil biodiversity and ecosystem function (Liu et al. 2024b, c, d).

Another key concept of NMDS is its ability to handle data from multiple sources and variables simultaneously. In soil science, it's common to measure various environmental factors alongside biological data, such as soil temperature, moisture, and nutrient levels. NMDS can incorporate these additional variables into the analysis, helping to reveal how they interact with fungal community composition. For example, you might find that moisture levels, in addition to pH, are also significant in shaping fungal communities. This multidimensional approach provides a more comprehensive understanding of the factors influencing soil ecosystems.

It's important to note that while NMDS is a powerful tool, it does not provide direct causal relationships between variables. Instead, it highlights patterns and relationships that can guide further investigation. For instance, if NMDS reveals a strong association between certain fungi and specific soil conditions, you might then design experiments to test how those conditions influence fungal growth and interactions. NMDS is best used as a starting point for exploring complex ecological data, generating hypotheses that can be tested through more targeted studies (Mahfouzi et al. 2024).

For young scientists in soil science, mastering NMDS opens up new possibilities for analyzing and interpreting ecological data. The ability to visualize complex datasets in a meaningful way is essential for advancing our understanding of soil biodiversity and the factors that shape it. As you gain experience with NMDS, you'll find that it's a versatile tool that can be applied to a wide range of ecological questions, from understanding the effects of environmental change on soil communities to monitoring the impact of conservation practices on soil health. NMDS is a valuable method for exploring the diversity and composition of soil fungi communities. By focusing on the rank order of differences between samples, NMDS provides a flexible and powerful way to visualize and interpret complex ecological data. Whether you are studying the effects of pH, land use, or other environmental factors on soil fungi, NMDS can help you uncover patterns and relationships that enhance our understanding of soil ecosystems and inform efforts to protect and manage these vital resources (Reis et al. 2024).

Main Assumptions Required for NMDS

The primary assumption required for Non-metric Multidimensional Scaling (NMDS) is that the rank order of dissimilarities among samples accurately reflects the underlying ecological relationships. Unlike methods that assume linear relationships or specific data distributions, NMDS focuses on preserving the order of dissimilarities, not the exact distances between samples. This means that NMDS assumes that the relative positions of samples in a low-dimensional space can represent the structure of the data based on these ranked dissimilarities. In the field of soil science, when studying the dynamics of soil microorganisms, this assumption allows researchers to focus on how similar or different microbial communities are across various samples, without being constrained by the need for linearity or normality in the data (Serrano et al. 2024).

Another key assumption in NMDS is that the dissimilarity metric chosen for the analysis appropriately captures the ecological differences between samples. In soil science, researchers often use metrics like Bray-Curtis dissimilarity or Euclidean distance to measure the differences in microbial community composition between soil samples. These metrics are sensitive to the presence or absence of species, as well as their relative abundances, which are critical factors in understanding microbial dynamics. The assumption here is that the chosen metric will accurately reflect the ecological distances between samples, allowing NMDS to create a meaningful ordination plot. For example, if you're comparing soil samples from different agricultural practices, the Bray-Curtis metric might be used to highlight differences in microbial communities caused by varying levels of tillage or fertilizer application (Shao et al. 2024).

NMDS also assumes that the dimensionality chosen for the analysis—usually two or three dimensions—is sufficient to capture the main patterns in the data without oversimplifying or losing important information. In practice, this means that the stress value, which measures the mismatch between the rank order of dissimilarities in the original high-dimensional space and the reduced low-dimensional space, should be low enough to indicate a good fit. In soil science, when studying microbial dynamics, a low stress value suggests that the NMDS plot accurately represents the relationships between different soil samples based on their microbial communities. For example, if you were studying the effects of different soil amendments on microbial composition, you would want the NMDS plot to clearly separate the samples treated with compost from those treated with synthetic fertilizers, assuming the stress value indicates a good fit (Wang et al. 2024b).

A related assumption is that the ecological processes influencing the microbial communities are adequately captured by the dissimilarities between samples. NMDS assumes that the rank order of these dissimilarities reflects the underlying ecological gradients or factors driving the differences in microbial composition. In soil science, this might involve assuming that factors like soil pH, organic matter content, or moisture levels are the primary drivers of microbial community dynamics, and that these factors are well-represented in the dissimilarities used for NMDS. For instance, if you're analyzing soil samples from a gradient of pH levels, NMDS assumes that the pH gradient is a key factor influencing the microbial communities and that the ordination plot will reflect this gradient. Another important assumption in NMDS is that the data used for the analysis are complete and representative of the microbial communities in the soil samples. NMDS can be sensitive to missing data or outliers, which can distort the ordination plot. In soil science, this means that the microbial community data should be collected systematically and thoroughly, ensuring that all relevant species are accounted for in each sample. For example, if certain soil samples have incomplete data due to technical issues during DNA sequencing, the resulting NMDS plot might not accurately reflect the true microbial dynamics, leading to misleading conclusions about the effects of soil management practices (Yu et al. 2024).

NMDS also assumes that the ecological relationships among samples are non-linear, making it a more appropriate choice than linear methods like Principal Component Analysis (PCA) in many cases. This assumption is particularly relevant in soil science, where microbial interactions and environmental gradients often exhibit non-linear patterns. For instance, the relationship between soil moisture and microbial activity is typically non-linear, with microbial communities responding differently to moisture levels depending on other factors like temperature or organic matter content. NMDS allows researchers to capture these complex, non-linear relationships and represent them in a way that reflects the true ecological dynamics of the system (Zhang et al. 2024a).

Lastly, NMDS assumes that the interpretation of the ordination plot will focus on the relative positions of samples rather than the specific axes. In NMDS, the axes are arbitrary and do not correspond to specific ecological gradients or variables, unlike in PCA where the axes represent principal components. This means that when interpreting NMDS results in soil science, the emphasis should be on understanding the overall patterns and relationships among samples, rather than on assigning specific meanings to the axes. For example, in a study of microbial dynamics in soils with different levels of organic matter, the key insight from NMDS would be the clustering of samples with similar microbial communities, rather than the position of these clusters along a specific axis. This approach allows for a more flexible interpretation of complex ecological data, which is particularly valuable in the study of soil microbial communities (Zhang et al. 2024b).

Steps to Conduct NMDS

Conducting Non-metric Multidimensional Scaling (NMDS) involves several critical steps, each of which plays a crucial role in ensuring that the analysis accurately reflects the underlying ecological patterns. For young scientists in soil science and soil microbiology, understanding these steps is essential for effectively applying NMDS to study the complex interactions between soil biota, environmental changes, and land use in organic soils. By following these steps, researchers can reveal important insights into how soil microbial communities respond to different environmental conditions and management practices (Xu et al. 2024).

The first step in conducting NMDS is data collection and preparation. This involves gathering data on soil biota, such as bacteria, fungi, and other microorganisms, from different sites that vary in terms of environmental conditions and land use. For example, you might collect soil samples from organic farms, natural forests, and disturbed lands to compare how microbial communities differ across these land uses. It's crucial to ensure that the data collected are comprehensive, representing the diversity of soil biota across all samples. After collection, the data must be cleaned and organized, with careful attention to standardizing measurements and dealing with any missing data (Botha et al. 2024).

Next, you need to choose an appropriate dissimilarity metric to quantify the differences between the samples. In soil microbiology, the Bray-Curtis dissimilarity is often used because it accounts for both the presence and abundance of species, making it particularly suitable for ecological data where species richness and evenness are important. For instance, if you're comparing microbial communities across organic soils under different land uses, the Bray-Curtis metric will help you assess how similar or different these communities are based on their composition and relative abundance. Selecting the right metric is crucial, as it will influence the results of the NMDS and how well it reflects the ecological relationships in your data (Chepkemoi et al. 2024).

Once you have chosen a dissimilarity metric, the third step is to compute the dissimilarity matrix, which is a table that shows the pairwise dissimilarities between all the samples. This matrix serves as the foundation for the NMDS analysis, providing the information needed to reduce the data to a lower-dimensional space. In the context of soil science, this matrix might reveal, for example, that microbial communities in organic soils from undisturbed forests are more similar to each other than to those from agricultural fields. The dissimilarity matrix is a crucial intermediary step that translates your raw data into a format that can be used for NMDS (Gómez et al. 2024).

The fourth step is to determine the number of dimensions for the NMDS ordination. NMDS typically reduces data to two or three dimensions, but the exact number should be chosen based on the stress value, which indicates how well the reduced dimensions represent the original dissimilarities. A lower stress value suggests a better fit. For example, in a study of soil biota under different land uses, you might find that a two-dimensional solution provides a sufficiently low stress value, making it a good choice for your NMDS plot. The goal is to balance simplicity with accuracy, choosing the fewest dimensions that still capture the main patterns in the data.

After determining the number of dimensions, the fifth step is to perform the NMDS ordination, which involves iteratively positioning the samples in the chosen dimensional space to best preserve the rank order of dissimilarities. The result is an NMDS plot where each point represents a soil sample, and the distances between points reflect the similarities or differences in microbial community composition. For example, in an NMDS plot of organic soils, you might see samples from agricultural fields clustering separately from those in natural forests, indicating distinct microbial communities shaped by different land uses. This visual representation is a powerful tool for interpreting complex ecological data (Idbella et al. 2024).

The sixth step is to assess the fit of the NMDS solution by examining the stress value and conducting goodness-of-fit tests. A stress value below 0.2 is generally considered acceptable, though lower values are preferred. If the stress is too high, it suggests that the NMDS plot may not accurately represent the data, and you might need to reconsider the number of dimensions or the dissimilarity metric used. In soil microbiology, a high stress value could mean that the relationships between microbial communities are too complex to be adequately captured in two or three dimensions, prompting further investigation. This step ensures that the NMDS results are reliable and meaningful.

Once you have a satisfactory NMDS solution, the seventh step is to interpret the ordination plot in the context of your research questions. This involves identifying patterns, such as clusters or gradients, and linking them to environmental factors or land use types. For instance, in an NMDS analysis of soil biota in organic soils, you might observe a clear separation between samples from areas with high organic matter content and those with lower levels, suggesting that organic matter is a key driver of microbial community composition. Interpretation should be done carefully, considering both the statistical results and the ecological context of the study.

The eighth step is to overlay environmental variables onto the NMDS plot to help explain the observed patterns. This can be done using techniques like vector fitting, which adds arrows representing environmental gradients to the plot. In soil science, you might overlay variables such as soil pH, moisture, or nutrient levels to see how they correlate with the microbial community composition. For example, you might find that soil pH strongly influences the distribution of fungal species across different land uses in organic soils. These overlays help to link the NMDS results with specific environmental factors, providing deeper insights into the dynamics of soil ecosystems.

Finally, the ninth step is to validate and communicate your findings. Validation can involve comparing the NMDS results with those from other analyses, such as redundancy analysis (RDA) or hierarchical clustering, to ensure consistency. Communicating your findings involves presenting the NMDS plot and your interpretations in a clear and accessible way, often with the aid of visualizations that highlight key patterns. For young scientists, it's essential to explain how NMDS helped uncover relationships between soil biota, environmental changes, and land use in organic soils, making the complex interactions within soil ecosystems easier to understand and manage (Huusko et al. 2024).

Handling High-Dimensional Soil Data with NMDS

Non-metric Multidimensional Scaling (NMDS) is particularly well-suited for handling high-dimensional data, which is common in soil science, especially when studying the complex communities of organisms like arbuscular mycorrhizal fungi (AMF). These fungi form symbiotic relationships with plants and are critical for nutrient cycling and soil health. High-dimensional data in this context might include various environmental parameters and the abundance or presence of multiple fungal species across many samples (Deng et al. 2024).

NMDS handles high-dimensional soil data by reducing the complexity of the data into a lower-dimensional space, typically two or three dimensions, while preserving the rank order of dissimilarities between samples. This process allows researchers to visualize patterns and relationships that might be difficult to detect in the original high-dimensional space. For example, when studying the composition of AMF communities across different soil types, you might collect data on numerous fungal species along with various environmental variables like soil pH, phosphorus levels, and organic matter content. Each sample would have a high-dimensional profile based on these variables, making direct analysis and visualization challenging (Liu et al. 2024a, d).

The first step in NMDS is to calculate the dissimilarity matrix, which captures the differences between each pair of samples based on the high-dimensional data. In the case of AMF, this could involve using the Bray-Curtis dissimilarity metric, which considers both the presence/absence and abundance of fungal species across the samples. The resulting dissimilarity matrix represents the high-dimensional data in a format that NMDS can work with, highlighting the relative similarities and differences between samples.

Next, NMDS iteratively places the samples in a lower-dimensional space, aiming to preserve the rank order of the dissimilarities from the original high-dimensional data. This means that samples with similar AMF communities will be positioned close to each other in the NMDS plot, while those with different communities will be farther apart. For instance, if you are comparing AMF communities in soils from forested areas, agricultural fields, and urban green spaces, NMDS might reveal distinct clusters corresponding to each land use type, reflecting the influence of environmental conditions on AMF diversity (Luo et al. 2024).

One of the key advantages of NMDS in handling high-dimensional soil data is that it doesn't assume linear relationships among variables or require the data to follow a normal distribution. This is particularly important in soil science, where interactions between AMF and environmental factors like soil pH or nutrient levels can be non-linear and complex. By focusing on rank orders rather than exact distances, NMDS provides a more flexible and accurate representation of these ecological relationships. For example, suppose you find that AMF communities in acidic soils with high organic matter content are quite similar, but those in neutral or alkaline soils with lower organic content are more varied. NMDS would allow you to visualize this pattern, even if the relationship between soil pH and AMF diversity is non-linear. The resulting ordination plot might show a gradient of samples along one axis, reflecting the influence of pH on AMF community structure, with clusters of samples indicating similar environmental conditions or management practices.

Another strength of NMDS in handling high-dimensional data is its ability to incorporate and overlay additional environmental variables onto the ordination plot. After performing the NMDS analysis, you can use vector fitting to add arrows representing gradients of soil properties like pH, phosphorus availability, or moisture content. This overlay helps interpret the NMDS plot by linking the observed patterns in AMF communities to specific environmental factors, providing insights into how these fungi respond to different soil conditions. For instance, you might overlay a vector representing soil phosphorus levels and observe that samples with higher phosphorus content tend to cluster together, suggesting that phosphorus is a significant factor influencing AMF community composition. This kind of analysis is invaluable for understanding the role of AMF in nutrient cycling and for developing soil management practices that promote healthy fungal communities.

In summary, NMDS effectively handles high-dimensional soil data by reducing its complexity into a more manageable form while preserving the essential ecological relationships. In the field of soil science, particularly when studying AMF, NMDS enables researchers to uncover and visualize patterns in fungal community composition across different environmental gradients and land use types, offering deep insights into the dynamics of these vital soil organisms (Mondal et al. 2024).

Why NMDS Is Particularly Useful in Soil Science Research?

Non-metric Multidimensional Scaling (NMDS) is particularly useful in soil science research because it provides a flexible and powerful method for analyzing complex, multidimensional data, which is common in studies of soil ecosystems. Soil environments are inherently variable, with numerous factors such as chemical properties, microbial communities, and physical structure interacting in non-linear ways. NMDS excels in capturing these complex relationships, making it an ideal tool for researchers studying soil, particularly in agricultural contexts where soils are heavily influenced by human activities (Reis et al. 2024).

One of the main advantages of NMDS in soil science is its ability to handle non-linear relationships and non-normal data distributions. Soil data often include variables like nutrient levels, organic matter content, pH, and microbial diversity, which do not necessarily follow simple linear patterns. For example, the relationship between soil organic matter and crop yield can be non-linear, with diminishing returns at higher organic matter levels. NMDS does not require data to fit a specific distribution or linear model, allowing it to accurately represent the complexities of soil data without forcing it into a simplified framework. In agricultural soils, NMDS can be particularly useful for studying the impacts of different farming practices on soil health. For instance, researchers might use NMDS to compare the microbial communities in soils managed under conventional tillage versus no-till systems. By analyzing the composition of microbial communities using NMDS, researchers can visualize how different management practices influence soil biodiversity. The resulting NMDS plot might show distinct clusters of soil samples from no-till fields, indicating a more stable and diverse microbial community compared to conventionally tilled soils, where frequent disturbance can lead to reduced microbial diversity (Ni et al. 2024).

NMDS is also valuable for integrating multiple environmental variables into a single analysis, providing a holistic view of soil health and functioning. Agricultural soils are influenced by a variety of factors, including fertilizer application, irrigation, and crop rotation. NMDS can incorporate these variables alongside biological data, such as the diversity and abundance of soil microbes, to reveal how these factors interact. For example, NMDS might show that soils with high nitrogen fertilizer inputs have microbial communities that differ significantly from those in organically managed soils, reflecting the impact of chemical inputs on soil ecology (Shao et al. 2024).

The visual output of NMDS, typically an ordination plot, is a powerful tool for interpreting and communicating complex soil data. In agricultural research, NMDS plots can help identify patterns that are not immediately apparent from raw data. For example, when comparing soil samples from different crop rotations, NMDS might reveal that certain rotations promote similar microbial communities, while others lead to divergent community structures. This insight can inform decisions about crop management practices that support soil biodiversity and long-term soil health. Moreover, NMDS is particularly useful for studying soil responses to environmental changes, such as those caused by climate change or shifts in agricultural practices. Agricultural soils are often subject to changes in temperature, moisture, and other conditions that affect soil properties and microbial dynamics. NMDS can help researchers track these changes over time by analyzing how soil properties and communities shift in response to environmental fluctuations. For instance, NMDS might be used to monitor how drought conditions affect soil microbial diversity across different types of agricultural soils, revealing which soils are more resilient to stress (Souza et al. 2024).

Another key application of NMDS in soil science is in the assessment of soil contamination and remediation efforts. Agricultural soils can be contaminated by heavy metals, pesticides, or other pollutants, which can significantly alter soil chemistry and biology. NMDS can be employed to study how these contaminants affect soil microbial communities and overall soil health. For example, an NMDS analysis might show that soils contaminated with heavy metals have microbial communities that are distinctly different from those in uncontaminated soils, indicating a disruption of normal soil processes. This information is crucial for developing effective remediation strategies that restore soil health (Wang et al. 2024a).

In studies of soil fertility, NMDS can be used to explore the relationships between soil properties and crop productivity. By analyzing soil data alongside crop yield data, NMDS can help identify which soil characteristics are most strongly associated with high yields. For instance, NMDS might reveal that soils with balanced nutrient levels and high microbial diversity tend to support higher crop yields, providing evidence for the importance of maintaining soil health through sustainable management practices. This type of analysis is particularly relevant for optimizing agricultural inputs to enhance productivity without degrading soil resources (Serrano et al. 2024).

NMDS also supports the exploration of soil heterogeneity within agricultural fields, which is important for precision agriculture. Agricultural fields are rarely uniform, with variations in soil properties that can affect crop growth. NMDS can be used to analyze soil samples from different parts of a field, helping to identify areas with distinct soil characteristics. For example, an NMDS analysis might show that certain areas of a field have soil properties that are more conducive to specific crops, allowing farmers to tailor their management practices to optimize productivity across the field.

Finally, NMDS is an invaluable tool for studying the effects of long-term agricultural practices on soil sustainability. Agricultural soils that have been managed for decades or centuries can exhibit complex changes in structure, chemistry, and biology. NMDS can help researchers untangle these long-term effects by analyzing how soil properties have evolved under different management regimes. For example, an

NMDS analysis of soils from long-term agricultural experiments might show how continuous monocropping has led to a decline in soil biodiversity compared to diversified cropping systems, providing insights into sustainable agricultural practices that maintain soil health over the long term.

Practical Application of NMDS Using Geostatistics in the Context of Biological Invasion

Applying Non-metric Multidimensional Scaling (NMDS) in the context of biological invasions allows researchers to analyze the complex interactions between native and invasive species within an ecosystem. When coupled with geostatistics, NMDS becomes an even more powerful tool, enabling scientists to assess spatial patterns and understand how invasive species spread and affect native biodiversity. This combination provides a practical approach for studying biological invasions, offering insights that can guide management and conservation efforts (Ni et al. 2024).

In a typical study of biological invasions, researchers might collect data on various species across different locations within an invaded landscape. These data include both the presence and abundance of native and invasive species, as well as environmental variables like soil properties, temperature, and moisture levels. NMDS is then used to reduce the complexity of this high-dimensional data, creating an ordination plot where each point represents a sampling location, and the distances between points reflect the ecological dissimilarities between sites. This plot helps visualize the impact of the invasive species on the local biodiversity, showing how different sites cluster based on species composition.

Geostatistics, on the other hand, provides the tools to analyze the spatial distribution of these ecological dissimilarities. By applying geostatistical methods like kriging, researchers can create spatial maps that predict the distribution of invasive species and their impact across the landscape. For example, after using NMDS to identify areas with similar species compositions, geostatistics can help interpolate these findings across the entire study area, highlighting hotspots of invasion or regions where native species are most at risk. This spatial information is crucial for targeting management efforts in the most affected areas.

In a practical application, consider a study of invasive plant species in a forested region. Researchers could sample vegetation across multiple plots, recording both the native species and the invasive plants. NMDS would be used to ordain these plots based on their species composition, revealing how the invasion has altered the plant communities. The ordination plot might show distinct clusters of plots where the invasive species dominate, indicating areas where native biodiversity has been most heavily impacted. By overlaying environmental variables like soil pH or moisture on the NMDS plot, researchers can further explore the conditions that facilitate or hinder the invasion (Souza et al. 2024).

Once the NMDS analysis is complete, the next step is to apply geostatistical methods to understand the spatial extent of the invasion. Using data from the NMDS plot, researchers can employ kriging to create a spatial prediction model of the invasive species' spread. This model would show how the invasion progresses across the landscape, potentially identifying corridors or barriers that influence the spread of the invasive species. For instance, geostatistics might reveal that the invasive plant species are more likely to spread along areas with disturbed soil, such as near roads or trails, providing valuable insights for land managers.

In addition to predicting the current distribution of invasive species, geostatistics can also be used to forecast future invasion scenarios. By incorporating temporal data into the NMDS and geostatistical models, researchers can simulate how the invasion might progress under different environmental conditions or management strategies. For example, a model might predict that increasing rainfall due to climate change could accelerate the spread of the invasive species, or that targeted removal efforts could slow the invasion in certain areas. These predictions help managers plan more effective interventions to control the invasion and protect native ecosystems (Shao et al. 2024).

The combination of NMDS and geostatistics is also valuable for assessing the effectiveness of management strategies aimed at controlling biological invasions. After implementing a management intervention, such as the removal of invasive species, researchers can use NMDS to compare the species composition of managed and unmanaged plots. If the NMDS analysis shows that the managed plots are moving closer to the composition of undisturbed native plots, this indicates that the management efforts are successful. Geostatistics can then be used to map the areas of recovery, providing a spatial assessment of the intervention's effectiveness.

Furthermore, this approach is not limited to plant invasions; it can also be applied to studies of invasive animals, microbes, or other organisms. For example, in studying the invasion of an exotic earthworm species into a forest, NMDS could be used to analyze how the earthworm affects soil microbial communities across different sites. Geostatistics would then allow researchers to map the spread of the earthworm and its impact on soil health, identifying areas where the invasion has caused significant changes in microbial diversity and function. This information is critical for understanding the broader ecological impacts of biological invasions. The integration of NMDS and geostatistics provides a robust framework for communicating the results of invasion studies to stakeholders, such as land managers, policymakers, and the public. The NMDS plots and spatial maps generated through geostatistics offer clear, visual representations of the invasion's impact and spread, making it easier to convey complex scientific findings. These tools can help stakeholders understand the urgency of the situation and the need for timely management actions to mitigate the effects of invasive species on native ecosystems (Luo et al. 2024).

Applying NMDS in Soil Science

Applying Non-metric Multidimensional Scaling (NMDS) in soil science offers young scientists a powerful way to explore and visualize complex ecological data. Soil ecosystems are intricate, with numerous interacting factors, such as microbial communities, chemical properties, and plant-soil relationships. NMDS is particularly useful in these contexts because it reduces high-dimensional data into a lower-dimensional space, making it easier to identify patterns and relationships that are not immediately apparent in the raw data. Understanding how to apply NMDS in soil science can significantly enhance your ability to interpret and communicate the findings from your research (Liu et al. 2024b).

The first step in applying NMDS is data collection, which often involves gathering a wide range of soil variables. For example, in soil microbiology, you might collect data on bacterial and fungal species composition across different soil samples, as well as environmental variables such as pH, moisture, and nutrient content. In plant production studies, you might include data on plant growth, nutrient uptake, and soil fertility. The key is to collect a comprehensive set of variables that capture the complexity of the soil ecosystem you are studying (Gómez et al. 2024).

Once you have your data, the next step is to choose an appropriate dissimilarity metric, which quantifies the differences between your samples. The Bray-Curtis dissimilarity is commonly used in soil science because it accounts for both the presence/absence and abundance of species, making it well-suited for ecological data. For instance, if you are studying the microbial communities in soils from different agricultural practices, the Bray-Curtis metric can help you assess how similar or different these communities are, based on the species present and their relative abundances (Huusko et al. 2024).

After calculating the dissimilarity matrix, you perform the NMDS ordination, which involves positioning your samples in a lower-dimensional space while preserving the rank order of dissimilarities. The result is an NMDS plot where each point represents a soil sample, and the distances between points reflect the ecological dissimilarities between them. For example, in a study of soil microbial communities under different cropping systems, NMDS might reveal distinct clusters of samples corresponding to each cropping system, indicating that different management practices lead to different microbial community structures (Botha et al. 2024).

One of the strengths of NMDS is its ability to handle non-linear relationships and non-normal data distributions, which are common in soil science. Soil properties and microbial communities often interact in complex, non-linear ways, making it difficult to model these relationships using traditional statistical methods. NMDS does not require your data to fit a specific distribution or model, allowing it to accurately capture the complexity of soil ecosystems. This flexibility makes NMDS particularly valuable for studying the effects of soil management practices, environmental changes, and other factors on soil health and productivity (Chen et al. 2024).

In soil microbiology, NMDS can be used to explore the diversity and composition of microbial communities in different soils. For example, if you are studying soils from organic and conventional farms, NMDS can help you visualize how the microbial communities differ between these farming systems. The NMDS plot might show that soils from organic farms cluster together, indicating more similar microbial communities, while soils from conventional farms are more dispersed, reflecting greater variability in microbial composition. This type of analysis can provide insights into how farming practices influence soil biodiversity and ecosystem functioning (Deng et al. 2024).

In plant production, NMDS can be applied to study the relationships between soil properties, plant growth, and crop yield. By analyzing soil and plant data together, NMDS can help you identify which soil characteristics are most strongly associated with high yields or healthy plant growth. For example, an NMDS analysis might reveal that soils with high levels of organic matter and balanced nutrient content are associated with more robust plant growth and higher yields. These findings can inform sustainable soil management practices that enhance productivity while maintaining soil health (Idbella et al. 2024).

Another important application of NMDS in soil science is in the assessment of soil contamination and remediation efforts. Agricultural soils can be contaminated by heavy metals, pesticides, or other pollutants, which can significantly alter soil chemistry and biology. NMDS can be used to study how these contaminants affect soil microbial communities and overall soil health. For example, an NMDS analysis might show that soils contaminated with heavy metals have microbial communities that are distinctly different from those in uncontaminated soils, indicating a disruption of normal soil processes. This information is crucial for developing effective remediation strategies that restore soil health.

NMDS is also useful for studying soil responses to environmental changes, such as those caused by climate change, land use changes, or agricultural practices. Soils are dynamic systems that respond to a wide range of environmental factors, and NMDS can help track these changes over time. For instance, you could use NMDS to analyze how soil microbial communities shift in response to drought or changes in land management. The NMDS plot might show that certain microbial communities are more resilient to stress, while others are more sensitive, providing valuable insights into the factors that influence soil resilience.

An additional advantage of NMDS is its ability to incorporate and overlay environmental variables onto the ordination plot. This allows you to link the observed patterns in your data to specific environmental factors, providing a deeper understanding of the underlying ecological processes. For example, you might overlay vectors representing soil pH, moisture, or nutrient levels onto your NMDS plot, revealing how these variables correlate with microbial community composition. This can help you identify the key drivers of soil biodiversity and function, which is essential for developing effective soil management strategies. NMDS is a powerful tool for communicating complex soil data to a broader audience, including scientists, policymakers, and land managers. The visual nature of NMDS plots makes it easier to convey the main findings of your research, highlighting the key patterns and relationships in your data. Whether you are studying the effects of farming practices on soil health, the impact of environmental changes on microbial communities, or the success of soil remediation efforts, NMDS provides a clear and intuitive way to present your results, making it an invaluable tool for soil scientists, microbiologists, and plant production experts alike (Liu et al. 2024b).

Software for NMDS

Several software packages are widely used for performing Non-metric Multidimensional Scaling (NMDS), each offering different features and capabilities. Here's an overview of some popular options:

1. R (vegan package): R is a powerful statistical software environment that includes a range of packages for NMDS analysis. The vegan package is particularly popular for ecological data analysis, including NMDS. It provides functions for performing NMDS, visualizing results, and conducting associated statistical analyses. To use NMDS in R, you would typically employ functions like metaMDS() for the ordination and plot() to visualize the NMDS results. The vegan package is well-documented and supported by a large user community, making it a versatile choice for NMDS.
2. MATLAB: MATLAB is another robust environment for numerical computing and visualization. While it does not have built-in NMDS functions, users can implement NMDS algorithms using custom scripts or available toolboxes. MATLAB's strength lies in its flexibility and the ability to handle complex data manipulation and visualization tasks. Users often write their own NMDS algorithms or adapt existing ones to fit their specific research needs.
3. PRIMER-E: PRIMER-E is a software package designed specifically for ecological and environmental data analysis. It includes tools for NMDS as well as other ordination and multivariate statistical methods. PRIMER-E provides a user-friendly interface and is well-suited for those who prefer graphical tools over programming. It also includes functions for visualizing NMDS results and interpreting the patterns in ecological data.
4. Canoco: Canoco is a software package tailored for multivariate data analysis in ecology. It offers a range of ordination methods, including NMDS, and is known for its robust statistical capabilities and ease of use. Canoco provides comprehensive graphical outputs and is often used for analyzing complex ecological data, making it suitable for detailed NMDS analyses and interpretations.
5. PCA/CA in Excel: For those who prefer working within a spreadsheet environment, there are add-ins for Excel that can perform Principal Component Analysis (PCA) and Correspondence Analysis (CA), which are related to NMDS. While Excel does not natively support NMDS, these add-ins can offer basic multivariate analysis capabilities. However, for more advanced NMDS functionality, specialized statistical software is recommended.
6. PC-ORD: PC-ORD is a software package designed for multivariate analysis of ecological data. It includes NMDS as one of its many features and is known for its user-friendly interface and extensive support for ecological and environmental data analysis. PC-ORD provides tools for ordination, clustering, and other multivariate methods, making it a comprehensive choice for NMDS applications.
7. TIBCO Spotfire: Spotfire is a data visualization and analytics software that supports a range of multivariate analysis techniques. Although it does not have a dedicated NMDS function, it allows for custom scripting and integration with R,

which means users can leverage R's NMDS capabilities within the Spotfire environment for advanced data analysis and visualization.

8. SYSTAT: SYSTAT is a statistical software package that includes various multivariate analysis methods. It offers tools for NMDS and provides a range of options for visualizing and interpreting results. SYSTAT is known for its user-friendly interface and integration of statistical methods, making it suitable for researchers who need to perform NMDS as part of a broader analysis.
9. Genuity: Genuity is a specialized software tool for ecological and environmental analysis, including NMDS. It provides a range of functions for ordination and multivariate statistics, with a focus on ecological data. Genuity is designed to handle complex datasets and provide detailed visualizations, making it a useful tool for NMDS applications in environmental research.

Example Application in Soil Science

Non-metric Multidimensional Scaling (NMDS) is a powerful analytical tool used in soil science to unravel complex relationships between soil properties, microbial communities, and plant interactions. One notable application is studying the impact of different agricultural practices on soil microbial diversity. Researchers can collect soil samples from fields managed under various practices—such as conventional tillage, no-till, and organic farming—and analyze the microbial communities present. NMDS can then be used to ordain these samples based on microbial composition, providing a visual representation of how different management practices influence soil biodiversity. The resulting NMDS plot might show clusters corresponding to each farming practice, revealing distinct microbial communities associated with each method.

Another application of NMDS is in assessing the effects of soil contamination on microbial communities. For instance, in a study examining soils contaminated with heavy metals, NMDS can help visualize how contamination alters microbial diversity and community structure. Researchers would first collect soil samples from contaminated and uncontaminated sites, then analyze the microbial communities using NMDS. The NMDS plot could reveal that microbial communities in contaminated soils are significantly different from those in uncontaminated soils, highlighting the impact of contamination on soil health. This information is crucial for developing remediation strategies to restore soil quality.

NMDS can also be employed to study the relationship between soil properties and plant growth. For example, researchers interested in understanding how variations in soil nutrient levels affect plant productivity might use NMDS to analyze soil samples with different nutrient profiles. By ordaining these samples based on their nutrient content and plant growth data, NMDS can reveal how specific nutrient combinations influence plant performance. An NMDS plot might show that certain nutrient levels are associated with higher plant growth, providing insights into optimal soil conditions for crop production.

In the context of soil degradation studies, NMDS can be used to assess the impact of land use changes on soil health. Researchers might compare soil samples from areas experiencing different degrees of degradation—such as deforestation, urbanization, and agricultural intensification. NMDS would help visualize how land use changes affect soil properties and microbial communities. For example, an NMDS analysis might show that soils from heavily degraded areas cluster together, indicating a loss of soil quality and biodiversity compared to less disturbed areas. This application helps identify critical areas needing restoration and informs land management practices.

Another significant application of NMDS is in evaluating the effects of soil management practices on soil carbon sequestration. Soil samples from fields with different management practices—such as cover cropping, reduced tillage, or organic amendments—can be analyzed using NMDS to determine how these practices influence soil carbon storage and microbial activity. The NMDS plot might reveal that certain practices lead to higher soil carbon levels and more diverse microbial communities, suggesting that these methods are more effective for enhancing soil health and mitigating climate change.

In studies of soil pH and its impact on soil ecosystems, NMDS can be used to analyze how variations in pH affect microbial community composition and soil fertility. Researchers might collect soil samples across a range of pH levels and use NMDS to visualize how pH influences microbial diversity and soil properties. An NMDS plot could show that soils with similar pH values cluster together, with distinct shifts in microbial communities and soil fertility observed as pH changes. This application helps understand how soil acidity affects ecosystem functions and guides pH management strategies.

NMDS is also useful for investigating the effects of soil texture on plant and microbial communities. Soil samples with varying textures—such as sandy, loamy, and clayey soils—can be analyzed to determine how texture influences soil properties and biological communities. By ordaining these samples using NMDS, researchers can visualize how soil texture affects microbial diversity and plant growth. The NMDS plot might reveal that different soil textures support distinct microbial communities and plant species, providing insights into the interactions between soil physical properties and biological functions.

Lastly, NMDS can be applied to study the impacts of climate change on soil ecosystems. Researchers might analyze soil samples from areas experiencing different climate conditions—such as varying temperatures or precipitation levels—and use NMDS to examine how these changes affect soil properties and microbial communities. The NMDS plot could highlight how climate-induced shifts in soil conditions influence microbial diversity and soil health, helping to predict how future climate scenarios might affect soil ecosystems. This application provides valuable information for developing adaptive management strategies to mitigate the effects of climate change on soil systems.

Conclusion

Non-metric Multidimensional Scaling (NMDS) stands out as a crucial tool in soil science for analyzing and interpreting complex, high-dimensional ecological data. Its strength lies in its ability to handle non-linear relationships and non-normal distributions, which are prevalent in soil studies involving microbial communities, soil properties, and plant interactions. NMDS provides a clear, visual representation of ecological patterns and relationships by reducing multidimensional data into a lower-dimensional space, making it easier to identify underlying trends and differences. This capability is particularly valuable for exploring the effects of soil management practices, environmental changes, and land use on soil health and biodiversity. By offering insights into how various factors influence soil ecosystems, NMDS enhances our understanding of soil dynamics and supports the development of informed management strategies aimed at improving soil quality and sustainability. As such, NMDS is an indispensable tool for researchers and practitioners in soil science, enabling them to derive meaningful conclusions from complex datasets and address critical questions in soil ecology and management.

References

Botha D, Barnard S, Claassens S et al (2024) Soil type and precipitation level have a greater influence on fungal than bacterial diversity in serpentine and non-serpentine biological soil crusts. Ecol Res. https://doi.org/10.1111/1440-1703.12500

Chen Y, Jiang Z, Ou J et al (2024) Nitrogen substitution practice improves soil quality of red soil (Ultisols) in South China by affecting soil properties and microbial community composition. Soil Tillage Res 240:106089. https://doi.org/10.1016/j.still.2024.106089

Chepkemoi L, Salifu D, Murungi LK, Tonnang HE (2024) High dimension multivariate data analysis for small group samples of chemical volatile profiles of African nightshade species. J Data Anal Infor Process 12:210–231. https://doi.org/10.4236/jdaip.2024.122012

Deng H, Ma X, Liu Z et al (2024) Soil ecosystem multifunctionality is strongly linked with crop yield after four decades chemical fertilization in black soil. Agric Ecosyst Environ 368:109007. https://doi.org/10.1016/j.agee.2024.109007

Dunleavy HR, Mack MC (2024) Nonlinear responses of ericaceous and ectomycorrhizal Arctic shrubs across a long-term experimental nutrient gradient. Ecosphere 15:e4888. https://doi.org/10.1002/ecs2.4888

Gómez AIP, Núñez-Rodríguez LX, Mora-Motta DA et al (2024) Unveiling soil bacterial diversity in the Andes-Amazon transition zone: impacts of forest conversion to pasture. Appl Soil Ecol 201:105486. https://doi.org/10.1016/j.apsoil.2024.105486

Huusko K, Manninen OH, Myrsky E, Stark S (2024) Soil fungal and bacterial communities reflect differently tundra vegetation state transitions and soil physico-chemical properties. New Phytol 243:407–422. https://doi.org/10.1111/nph.19808

Idbella M, Baronti S, Vaccari FP et al (2024) Long-term application of biochar mitigates negative plant–soil feedback by shaping arbuscular mycorrhizal fungi and fungal pathogens. Microorganisms 12:810. https://doi.org/10.3390/microorganisms12040810

Islam W, Zeng F, Almoallim HS, Ansari MJ (2024) Unveiling soil animal community dynamics beneath dominant shrub species in natural desert environment: implications for ecosystem management and conservation. J Environ Manag 366:121697. https://doi.org/10.1016/j.jenvman.2024.121697

Liu D, Fei YH, Peng Y et al (2024a) Genotype of pioneer plant Miscanthus is not a key factor in the structure of rhizosphere bacterial community in heavy metal polluted sites. J Hazard Mater 477:135242. https://doi.org/10.1016/j.jhazmat.2024.135242

Liu WS, Zhu ZH, Lu ZG et al (2024b) Soil quality associated with microbial community characteristics and dominant taxa across different tillage practices. Land Degrad Dev. https://doi.org/10.1002/ldr.5249

Liu X, Yu Y, Yu H et al (2024c) Nonbiodegradable microplastic types determine the diversity and structure of soil microbial communities: a meta-analysis. Environ Res 260:119663. https://doi.org/10.1016/j.envres.2024.119663

Liu Z, Nan Z, Lin S et al (2024d) Peanut-based intercropping systems altered soil bacterial communities, potential functions, and crop yield. PeerJ 12:e16907. https://doi.org/10.7717/peerj.16907

Lu H, Ma K, Chen X et al (2024) Multiple soil health assessment methods for evaluating effects of organic fertilization in farmland soil of agro-pastoral ecotone. Agriculture 14:572. https://doi.org/10.3390/agriculture14040572

Luo X, Zhang L, Li Z et al (2024) Adaptation mechanisms of soil microbial community to stoichiometric imbalances caused by forest conversion. Appl Soil Ecol 201:105529. https://doi.org/10.1016/j.apsoil.2024.105529

Maciel MGR, Ramos MB, de Souza SM et al (2024) Water and nutrients mediate tree communities in the driest region of Caatinga. Catena 242:108107. https://doi.org/10.1016/j.catena.2024.108107

Mahfouzi M, Hamidian AH, Kaboli M (2024) Edaphic drivers influencing forage grasslands in Bujagh National Park, Iran. Resources 13:29. https://doi.org/10.3390/resources13020029

Mondal S, Burgos-Hernandez T, Ralston TI et al (2024) Urban greenspaces shape soil nematode community across soil depth gradients: belowground life at The Ohio State University. Ecol Indic 166:112399. https://doi.org/10.1016/j.ecolind.2024.112399

Ni X, Zhu X, Feng Q et al (2024) Effect of application rates of N and P fertilizers on soil nematode community structure in Mollisols. Agronomy 14:507. https://doi.org/10.3390/agronomy14030507

Reis BR, Vasconcelos ALS, Silva AMM et al (2024) Changes in soil bacterial community structure in a short-term trial with different silicate rock powders. Chem Biol Technol Agric 11:61. https://doi.org/10.1186/s40538-024-00586-w

Serrano AR, Peco B, Morillo JA, Ochoa-Hueso R (2024) Abandonment of traditional livestock grazing reduces soil fertility and enzyme activity, alters soil microbial communities, and decouples microbial networks, with consequences for forage quality in Mediterranean grasslands. Agric Ecosyst Environ 366:108932. https://doi.org/10.1016/j.agee.2024.108932

Shao L, Yan P, Ye S et al (2024) Soil organic matter and water content affect the community characteristics of arbuscular mycorrhizal fungi in Helan mountain, an arid desert grassland area in China. Front Microbiol 15:1377763. https://doi.org/10.3389/fmicb.2024.1377763

Souza T, Dobner M Jr, Batista DS et al (2024) Site quality for Araucaria angustifolia plantations with subtropical Cambisol is driven by soil organism assemblage and the litter and soil compartments. Forests 15:510. https://doi.org/10.3390/f15030510

Wang M, Sui X, Wang X et al (2024a) Soil fungal community differences in manual plantation larch forest and natural larch forest in northeast China. Microorganisms 12:1322. https://doi.org/10.3390/microorganisms12071322

Wang W, Yang Y, Li J et al (2024b) Consecutive fertilization-promoted soil nutrient availability and altered rhizosphere bacterial and bulk fungal community composition. Forests 15:514. https://doi.org/10.3390/f15030514

Xu F, Li C, Chen Y et al (2024) Soil microbial community structure and soil fertility jointly regulate soil microbial residue carbon during the conversion from subtropical primary forest to plantations. Geoderma 441:116767. https://doi.org/10.1016/j.geoderma.2023.116767

Yu Y, Chen L, Duan W (2024) Responses of bacterial and fungal communities to short-term nitrogen and phosphorus additions in temperate forest soil aggregates in northeastern China. Appl Soil Ecol 197:105341. https://doi.org/10.1016/j.apsoil.2024.105341

Zhang J, Kang L, Cao Y, Zhang C, Wang QW (2024a) Litter decomposition in pure and mixed plantations on the Loess Plateau, China: lack of home-field advantage. Catena 244:108239. https://doi.org/10.1016/j.catena.2024.108239

Zhang Y, Chen Y, An B et al (2024b) The succession patterns and drivers of soil bacterial and fungal communities with stand development in Chinese fir plantations. Plant Soil 500:547–569. https://doi.org/10.1007/s11104-024-06502-3

Canonical Correspondence Analysis (CCA) 7

Abstract

Canonical Correspondence Analysis (CCA) is a multivariate statistical method used to explore the relationships between biological communities and environmental variables, aiming to identify how specific environmental gradients influence community composition. Unlike Principal Component Analysis (PCA), which focuses on reducing the dimensionality of data by capturing the maximum variance, CCA explicitly links environmental variables with species data, making it particularly suited for analyzing ecological data where the goal is to understand these specific relationships. The primary assumptions of CCA include linear relationships between variables, homogeneity of variance, and the assumption that data is appropriately scaled and transformed. Multiple CCA extends this approach to handle high-dimensional soil data by analyzing multiple sets of environmental and biological variables simultaneously, allowing for a comprehensive exploration of complex interactions. In soil science research, CCA is valuable for its ability to integrate and interpret multifaceted data, revealing how soil properties and ecological factors interact to shape soil health and ecosystem functions. To ensure reliable results, researchers should validate their data quality, check for multicollinearity among variables, and ensure that the assumptions of CCA are met, including proper data transformation and scaling.

Keywords

Ecological research · Environmental variables · Canonical correspondence analysis (CCA) · Ordination · Species composition

Questions Covered in the Chapter

(a) What is the primary difference between CCA and PCA?
(b) What are the main assumptions required for CCA?

T. Souza, *Advanced Statistical Analysis for Soil Scientists*,
https://doi.org/10.1007/978-3-031-88161-9_7

(c) How does CCA handle high-dimensional soil data?
(d) Why is CCA particularly useful in soil science research?
(e) What should researchers check for before applying CCA to ensure reliable results?

Introduction

Canonical Correspondence Analysis (CCA) is an advanced statistical technique widely used in ecological research to explore the relationships between species composition and environmental variables. Unlike traditional ordination methods, CCA directly relates species data to environmental factors, making it a valuable tool for understanding how different variables influence community structure. This method is particularly effective in studies where the goal is to identify the environmental gradients that drive species distributions. One of the core features of CCA is its ability to constrain ordination axes to be linear combinations of the measured environmental variables. This means that the analysis directly links species distribution patterns to specific environmental gradients, allowing for more interpretable results. The assumption that species respond linearly to environmental changes underpins the method, making it suitable for datasets where this relationship is expected to hold true (Bashtian et al. 2024).

Data for CCA typically consist of two matrices: one containing species abundance or presence/absence data across different sampling sites, and another detailing the environmental variables measured at these sites. The strength of CCA lies in its ability to simultaneously analyze these matrices, providing insights into how environmental factors influence species composition. This dual focus on species and environment is what sets CCA apart from other multivariate techniques. One of the main benefits of using CCA is its ability to produce results that are both statistically robust and ecologically meaningful. By directly incorporating environmental variables into the ordination process, CCA allows researchers to pinpoint which factors are most influential in shaping community composition. This makes it a preferred method for studies aiming to understand the ecological impacts of environmental changes (Kong et al. 2024).

The utility of CCA extends to various fields, including soil science, where it has been used to explore the complex interactions between soil properties and biological communities. By linking soil characteristics with the diversity and abundance of soil organisms, CCA helps researchers understand how different environmental factors contribute to the structure of these communities. This information can be crucial for developing strategies to maintain or improve soil health. The insights gained from CCA are particularly valuable for understanding how environmental conditions affect biological communities over time. This makes it an important tool for monitoring ecological changes, especially in response to human activities or environmental stressors. By identifying key environmental drivers, CCA can inform management practices aimed at conserving biodiversity and ensuring ecosystem resilience (Rivera-Zizumbo et al. 2024).

While CCA offers many advantages, it is not without its challenges. The assumption of linearity may not always be appropriate, especially in ecosystems where species responses to environmental gradients are nonlinear. Additionally, the method requires careful selection and accurate measurement of environmental variables, as these directly influence the interpretation of the results. Despite these challenges, CCA remains a powerful method for exploring the complex relationships between species and their environments. Its ability to handle large datasets and produce results that are both statistically valid and ecologically relevant makes it an essential tool for researchers. The insights it provides can guide efforts to preserve biodiversity and manage ecosystems in a sustainable way.

Canonical Correspondence Analysis is a valuable approach for studying species-environment relationships. Its direct integration of environmental variables into the analysis offers a clear advantage in ecological research. While it requires careful consideration of its assumptions and data requirements, the benefits of CCA in producing meaningful ecological insights make it an indispensable method in the study of ecological communities (Wang et al. 2024a).

Main Differences Between CCA and PCA

The primary difference between Canonical Correspondence Analysis (CCA) and Principal Component Analysis (PCA) lies in how they handle relationships between variables and the type of questions they answer. CCA is a constrained ordination method that specifically examines the relationship between two datasets, such as species data and environmental variables. In contrast, PCA is an unconstrained ordination method that focuses solely on finding the main patterns of variation within a single dataset, without reference to external variables. CCA explicitly links species composition to environmental factors. For example, in plant nutrition, CCA might be used to determine how soil nutrient levels influence the distribution and abundance of various plant species. The analysis seeks to identify which nutrients are most strongly associated with the presence of certain plants, providing direct insights into how these environmental factors shape community structure. This makes CCA particularly valuable when the research goal is to understand how external variables drive biological patterns (Tang et al. 2024).

On the other hand, PCA is concerned with reducing the dimensionality of a dataset by identifying the directions (principal components) along which the variation in the data is greatest. In soil fertility research, PCA might be applied to a dataset of soil properties (e.g., pH, organic matter, nutrient levels) to identify the main axes of variation. PCA reveals which combinations of these properties explain the most variance in the dataset, helping to simplify complex datasets into a few key variables. However, PCA does not provide direct information about how these variables might relate to external factors, such as plant species or environmental conditions. In CCA, the ordination axes are constrained to be linear combinations of the environmental variables. This means that the analysis is guided by the environmental data, producing results that are directly interpretable in the context of those variables. For instance, in a study on plant nutrition, CCA might reveal that phosphorus and potassium levels are the primary factors influencing plant community composition across different soil types. The resulting ordination would clearly show how plant species are distributed along these nutrient gradients (Zhu et al. 2024).

PCA, by contrast, identifies axes of variation based solely on the structure of the data itself, without any external constraints. In a soil fertility study, PCA might reveal that soil pH and organic matter content account for the most variation in soil properties. However, PCA does not explain why these variables are important or how they might relate to plant growth or other environmental factors. It simply identifies the main patterns within the dataset, leaving interpretation to the researcher. One of the strengths of CCA is its ability to produce ecologically meaningful results by directly linking species data with environmental variables. For example, if a researcher is studying the effect of soil nutrient levels on crop yield, CCA could help identify which nutrients are most strongly associated with high yields. This information could then be used to guide fertilizer application or other management practices (Zhang et al. 2024a).

The direct connection between environmental variables and biological outcomes is a key advantage of CCA over PCA. In contrast, PCA is often used when the primary goal is data exploration or simplification, rather than direct interpretation. In plant nutrition, PCA might be used to summarize a large number of soil properties into a few principal components that capture the majority of the variation. This can be useful for identifying underlying patterns or for reducing the complexity of a dataset before applying other statistical analyses. However, the results of PCA do not directly address the relationship between the soil properties and plant performance (Ng et al. 2024).

Another key difference is in the interpretation of the results. CCA provides ordination diagrams where species and environmental variables are plotted together, making it easy to see how the variables relate to each other. In the context of soil fertility, this might mean that plants preferring high nitrogen soils are clearly separated from those that thrive in low-nitrogen conditions. PCA, however, produces ordination diagrams that show how the variables within a single dataset relate to the principal components. This can reveal which variables are most correlated, but it does not provide direct information about external factors or species distribution. Finally, the choice between CCA and PCA depends on the research question. If the goal is to explore the relationships between species composition and environmental variables, CCA is the appropriate choice. For instance, in a study examining how soil nutrient gradients affect the diversity of a plant community, CCA would be used to directly relate nutrient levels to species distribution. Conversely, if the objective is to identify the main sources of variation in a set of soil properties, without reference to external factors, PCA would be the better tool (Kaewfoo et al. 2024).

Key Concepts of CCA

Canonical Correspondence Analysis (CCA) is a method that helps researchers understand how different environmental factors influence the composition of biological communities. In soil science and soil microbiology, CCA is particularly useful for examining how soil properties affect the distribution and abundance of organisms like fungi. By analyzing these relationships, CCA provides insights into the factors that shape microbial communities in various soil environments. Imagine a study where researchers want to explore how soil pH, nutrient levels, and moisture influence the diversity of fungal species in different soil samples. CCA allows them to link these soil properties directly to the composition of the fungal community, helping to identify which factors are most strongly associated with certain species. This approach offers a clear picture of how environmental gradients drive the distribution of organisms in the soil (Dayang Najwa et al. 2024).

One of the strengths of CCA is its ability to handle complex datasets where multiple environmental variables are involved. For example, in a study of soil chemical properties, CCA can simultaneously analyze the effects of pH, phosphorus, nitrogen, and other nutrients on fungal diversity. This comprehensive analysis helps researchers understand the combined influence of these factors on the community structure, revealing interactions that might not be apparent when examining each variable separately. Another important aspect of CCA is its ability to produce ordination plots that visually represent the relationships between species and environmental variables. In the context of soil microbiology, these plots might show how different fungal species are distributed along gradients of soil acidity or nutrient availability. Researchers can quickly see which species are associated with specific soil conditions, making the results easy to interpret and apply to real-world situations (Chernysheva and Fornasier 2024).

When using CCA, it's essential to carefully select the environmental variables included in the analysis. In soil science, this might involve choosing the most relevant soil properties that are likely to influence microbial communities, such as organic matter content, pH, or moisture levels. By focusing on the most impactful variables, researchers can ensure that the analysis provides meaningful insights into the factors driving community composition. CCA also helps in identifying the environmental conditions that favor or inhibit certain species. For instance, in a study on soil fungi, CCA might reveal that certain species thrive in high-nitrogen soils, while others are more abundant in acidic conditions. Understanding these preferences is crucial for managing soil health, as it allows for the optimization of conditions to support beneficial microbial communities (Ghafoor et al. 2024).

In addition to revealing direct relationships between environmental variables and species composition, CCA can uncover complex interactions between factors. For example, the combined effect of soil pH and moisture might have a more significant impact on fungal diversity than either factor alone. CCA helps to disentangle these interactions, providing a clearer understanding of how multiple variables work together to shape microbial communities. CCA is also valuable for exploring how changes in soil properties, due to factors like agricultural practices or environmental stressors, influence microbial communities over time. By comparing CCA results from different time points or locations, researchers can assess how shifts in soil conditions affect the composition of fungi and other organisms. This information is critical for developing strategies to maintain or improve soil health in the face of changing environmental conditions (Tenaw et al. 2024).

One of the practical applications of CCA in soil science is in guiding soil management practices. For example, if CCA identifies that high levels of a particular nutrient are associated with a decrease in beneficial fungi, land managers can adjust fertilization practices to avoid negative impacts on soil biodiversity. This application of CCA helps to inform decisions that promote sustainable soil management. The interpretation of CCA results requires an understanding of how the ordination plots relate to the underlying data. In soil microbiology, this might involve recognizing that species located close to certain environmental variables in the plot are more strongly associated with those conditions. By carefully interpreting these plots, researchers can draw meaningful conclusions about the relationships between soil properties and microbial communities (Kesbi et al. 2024).

Another important consideration when using CCA is the potential influence of outliers or unusual data points. In soil science research, outliers might represent rare or extreme soil conditions that could disproportionately affect the results. By identifying and addressing these outliers, researchers can ensure that the CCA provides a more accurate representation of the typical relationships in the data. Finally, CCA encourages young scientists to think critically about the interactions between environmental variables and biological communities. By applying CCA to studies of soil chemical properties and fungal composition, they can develop a deeper understanding of how soil environments influence microbial diversity. This knowledge is essential for addressing the challenges of soil management and conservation, as it equips researchers with the tools to promote healthy and productive soils (Neupert et al. 2024).

Main Assumptions Required for CCA

Canonical Correspondence Analysis (CCA) is a widely used technique for exploring the relationships between species composition and environmental variables. However, for the results of CCA to be reliable and meaningful, certain key assumptions must be met. These assumptions ensure that the method accurately reflects the ecological processes being studied, particularly in complex environments like degraded soils under climate change. One of the primary assumptions of CCA is that the relationships between species abundance and environmental variables are linear. This means that the species' response to environmental gradients, such as soil moisture or nutrient levels, is expected to change in a straight-line fashion. In the context of degraded soils affected by climate change, this assumption might be tested by examining how fungal communities respond to shifts in soil pH or organic matter content. If the response is nonlinear, the results of CCA could be misleading (Wang et al. 2024b).

Another crucial assumption is that species distributions are unimodal along environmental gradients. This implies that each species has an optimal range of environmental conditions, where its abundance peaks, and decreases as conditions move away from this optimum. For instance, in degraded soils with varying levels of soil salinity due to climate change, certain fungi might thrive at moderate salinity levels but decline in extremely high or low salinity conditions. CCA assumes this kind of bell-shaped distribution when analyzing species-environment relationships. The assumption of species-environment independence is also important. CCA assumes that the environmental variables included in the analysis are not strongly correlated with one another. In soil science, particularly in studies of degraded soils under climate change, variables such as soil moisture, temperature, and nutrient levels might be interrelated. If these variables are too closely correlated, it can lead to multicollinearity, which can distort the results of CCA and make it difficult to determine the true effects of each variable (Xing et al. 2024).

CCA also assumes that the environmental variables are accurately measured and represent the actual conditions that influence species composition. In the context of studying degraded soils affected by climate change, this means that researchers need to carefully measure variables like soil moisture, temperature, pH, and nutrient levels. Any inaccuracies or biases in these measurements can lead to incorrect interpretations of how these factors are driving changes in microbial communities (Zhang et al. 2024b).

The assumption of homogeneity in species-environment relationships across the study area is another critical factor. CCA assumes that the relationships between species and environmental variables are consistent throughout the study area. However, in degraded soils under the influence of climate change, these relationships might vary across different locations due to localized effects, such as differences in soil texture or human activities. If this heterogeneity is not accounted for, it could lead to misleading results when using CCA. It is also assumed that the species included in the analysis respond to the measured environmental variables and not to other unmeasured factors. In degraded soils, particularly those affected by climate change, unmeasured variables such as microclimatic conditions or historical land use might influence fungal communities. If these factors are not included in the analysis, the CCA results might not fully capture the true drivers of species composition changes (Zhou et al. 2024).

Another assumption is that the sampling sites are representative of the broader area being studied. In soil science research focused on degraded soils under climate change, it is essential to ensure that the sites selected for sampling adequately represent the range of conditions present in the larger landscape. If the sampling is biased towards certain conditions, the CCA results may not accurately reflect the broader patterns of species-environment relationships. The assumption of a sufficient sample size is also critical. CCA requires a relatively large number of sampling sites to provide robust and reliable results. In studies of degraded soils, particularly in areas impacted by climate change, it is important to collect data from a wide range of locations to capture the full spectrum of environmental variability and its effects on microbial communities. Insufficient sample sizes can lead to overfitting and unreliable conclusions (Zhang et al. 2024c).

CCA assumes that the species included in the analysis are ecologically relevant and responsive to the environmental gradients being studied. In the context of degraded soils under climate change, this means focusing on microbial species that are known to be sensitive to changes in soil conditions, such as fungi that are important for nutrient cycling or soil structure. Including irrelevant or insensitive species in the analysis can dilute the strength of the relationships being examined, leading to weaker and less informative results (Xing et al. 2024).

Steps to Conduct CCA

Conducting Canonical Correspondence Analysis (CCA) begins with clearly defining the research question, especially when studying soil science and the biological invasion of tree species. For instance, a researcher might be interested in understanding how soil properties like pH, moisture, and nutrient levels influence the spread of an invasive tree species in a particular ecosystem. This initial step involves identifying the specific environmental variables and species data that will be analyzed using CCA. The next step involves collecting and preparing the data. In the context of soil science and tree species invasion, this could involve gathering soil samples from different sites where the invasive species is present and where it is absent. Researchers would measure various soil properties, such as nutrient levels, pH, and moisture, as well as record the abundance or presence of the invasive species and other native vegetation. Ensuring that the data is accurate and representative of the study area is crucial for the reliability of the CCA results (Wang et al. 2024b).

Once the data is collected, it must be organized into a format suitable for CCA. Typically, this involves creating two matrices: one for the species data and one for the environmental variables. In the case of biological invasion, the species matrix would include the abundance or presence-absence data for the invasive tree species and other relevant species, while the environmental matrix would contain the corresponding soil properties for each site. Proper organization of these matrices is essential for the analysis to proceed correctly. Before running CCA, it's important to explore the data to check for any issues such as outliers, missing values, or multicollinearity among the environmental variables. For example, if soil pH and moisture are highly correlated, it might be necessary to address this before proceeding with the analysis. This step ensures that the CCA results will be robust and that the relationships identified between soil properties and species distribution are not artifacts of data issues (Rivera-Zizumbo et al. 2024).

The next step involves performing the CCA using statistical software. In this process, the software will relate the species data to the environmental variables, generating ordination plots that visually represent these relationships. For instance, the CCA might reveal that certain soil conditions, like low pH or high nitrogen levels, are strongly associated with the presence of the invasive tree species. These visualizations help researchers interpret the complex interactions between soil properties and species composition. After running the CCA, it is important to interpret the ordination plots. In these plots, sites, species, and environmental variables are represented by points or vectors. The proximity of a species to an environmental variable in the plot indicates a strong relationship. For example, if the invasive tree species clusters closely with high soil nitrogen levels, this suggests that nitrogen-rich soils may be facilitating its spread. Interpreting these plots accurately is key to drawing meaningful conclusions from the CCA.

Once the relationships have been interpreted, researchers should assess the statistical significance of the CCA results. This can involve performing permutation tests or other statistical tests to determine whether the observed relationships are likely to have occurred by chance. In the context of biological invasions, demonstrating that the association between certain soil properties and the invasive species is statistically significant adds credibility to the findings. The next step involves validating the CCA model. Researchers might do this by applying the CCA to a subset of the data or to a new dataset to see if similar patterns emerge. In the study of soil properties and invasive tree species, validation might involve testing whether the identified soil conditions consistently predict the presence of the invasive species in different locations or under different environmental conditions. Validation helps confirm the robustness of the CCA results (Tang et al. 2024).

After validation, the results should be interpreted in the broader ecological context. For example, if CCA reveals that the invasive tree species is strongly associated with disturbed soils high in nitrogen, researchers might explore how land use changes or nitrogen deposition are contributing to the invasion. This step connects the statistical findings to real-world ecological processes and can inform management strategies. Communicating the results of the CCA is an important step in the process. Researchers should present their findings clearly, using the ordination plots and statistical results to support their conclusions. In the context of soil science and biological invasions, this might involve discussing how specific soil management practices could mitigate the spread of the invasive tree species, based on the CCA findings (Kesbi et al. 2024).

The implications of the CCA results should be considered for future research and management. For instance, if certain soil conditions are identified as key drivers of the invasive species' success, future studies might focus on monitoring these conditions or testing interventions to alter them. In the broader context of ecosystem management, the CCA results could inform strategies to prevent the further spread of invasive species by managing soil properties. By following these steps, young scientists can effectively conduct CCA and gain valuable insights into the relationships between environmental variables and species composition. This knowledge is particularly important in fields like soil science and ecology, where understanding these relationships can help address challenges such as biological invasions and ecosystem degradation (Dayang Najwa et al. 2024).

Handling High-Dimensional Soil Data with Multiple CCA

Multiple Canonical Correspondence Analysis (multiple CCA) is an extension of traditional CCA that is particularly useful for handling high-dimensional soil data, where there are many environmental variables and species involved. This method is designed to explore the relationships between multiple sets of variables, making it an effective tool in complex ecological studies such as those involving the invasion of *Prosopis*, *Acacia*, and *Pinus* species in plant ecology. In the context of plant ecology and biological invasions, high-dimensional soil data might include a wide range of soil properties, such as pH, moisture, nutrient levels, and organic matter content, measured across different sites. These sites could be invaded by one or more species like *Prosopis*, *Acacia*, and *Pinus*. The challenge is to understand how these various soil properties influence the distribution and abundance of these invasive species, and how the species themselves interact with the soil environment.

Multiple CCA handles this complexity by allowing researchers to simultaneously analyze multiple environmental datasets against species composition data. For example, suppose a study is examining how the invasion of *Prosopis*, *Acacia*, and *Pinus* species is related to a combination of soil factors like nitrogen levels, phosphorus availability, and soil pH across different regions. In this case, multiple CCA can be used to analyze these soil variables together, providing a comprehensive view of how they collectively influence the presence and spread of these invasive species. One of the key advantages of multiple CCA in handling high-dimensional soil data is its ability to reduce the dimensionality of the data while retaining the most important relationships. In the example of *Prosopis*, *Acacia*, and *Pinus* invasions, the method might reveal that while many soil factors are measured, only a few (e.g., soil pH and nitrogen levels) are strongly associated with the distribution of these species. By focusing on the most relevant variables, multiple CCA simplifies the analysis and makes it easier to interpret the ecological relationships.

Another important feature of multiple CCA is its ability to handle multicollinearity, which is common in high-dimensional data where many environmental variables are often correlated. For instance, soil moisture and organic matter might be highly correlated in areas invaded by Acacia. Multiple CCA can partition the variance attributable to each variable, allowing researchers to determine the unique contribution of each soil factor to the presence of the invasive species. This helps in understanding which soil properties are truly driving the invasion. Multiple CCA also provides insights into how different invasive species might respond differently to the same soil conditions. In a study involving *Prosopis*, *Acacia*, and *Pinus* species, multiple CCA could reveal that while all three species are associated with certain soil properties, they might have different optimal conditions. For example, Prosopis might thrive in soils with higher pH and lower moisture, while *Pinus* prefers more acidic and nutrient-rich soils. This differentiation helps in understanding the ecological niches that each invasive species occupies.

Furthermore, multiple CCA can identify interaction effects between environmental variables and species composition. In the case of *Prosopis*, *Acacia*, and *Pinus* species, the analysis might show that the combined effect of soil nitrogen and pH is more influential in determining species distribution than either factor alone. This ability to model interactions is crucial in ecological studies, where the effects of multiple environmental factors often combine in complex ways to influence species invasions. In practical terms, multiple CCA can be used to inform management strategies aimed at controlling invasive species. For instance, if the analysis reveals that high nitrogen levels strongly promote the invasion of *Prosopis*, land managers might focus on reducing nitrogen inputs in vulnerable areas. Similarly, understanding that *Acacia* thrives in disturbed soils might lead to strategies that focus on soil restoration and reducing disturbance as a way to prevent further invasions.

Multiple CCA also allows for the comparison of different invaded sites, helping researchers to understand how the relationships between soil properties and species composition vary across regions. This can be particularly valuable in large-scale studies where *Prosopis*, *Acacia*, and *Pinus* are invading diverse ecosystems. By comparing sites with different invasion histories and soil conditions, multiple CCA can highlight the factors that make certain areas more susceptible to invasion. Finally, multiple CCA facilitates a more holistic understanding of the invasion process by integrating multiple datasets. For example, researchers might combine soil chemical properties, climatic data, and species composition to gain a comprehensive understanding of how these factors interact to drive the invasion of *Prosopis*, *Acacia*, and *Pinus* species. This integrated approach is essential for developing effective, science-based strategies to manage and mitigate the impacts of invasive species in plant ecology.

Why CCA Is Particularly Useful in Soil Science Research?

Canonical Correspondence Analysis (CCA) is particularly useful in soil science research because it allows for the simultaneous analysis of species composition and environmental variables, which is essential in understanding the complex interactions between soil properties and the organisms that inhabit them. Soil ecosystems are inherently complex, with numerous chemical, physical, and biological factors interacting in ways that influence the distribution and abundance of soil organisms, including plants, microbes, and invertebrates. CCA is uniquely suited to disentangle these relationships, making it a powerful tool in soil science and soil ecology (Bashtian et al. 2024).

One of the main reasons CCA is valuable in soil science is its ability to handle multiple, often correlated, environmental variables. For example, in a study of soil microbial communities, researchers might be interested in how factors like pH, moisture, organic matter, and nutrient levels influence the composition of bacteria or fungi in the soil. CCA can analyze these variables together, helping to identify which factors are most strongly associated with changes in microbial community structure. This is particularly useful in soils with varying properties, where the effects of individual variables might be difficult to isolate (Kaewfoo et al. 2024).

CCA is also particularly effective in studies that involve gradients of environmental conditions, such as those found in soil chronosequences or across different land uses. For instance, in a study of soil succession following agricultural abandonment, researchers might use CCA to explore how soil properties change over time and how these changes influence plant colonization and microbial community shifts. By ordinating sites along these gradients, CCA can reveal the trajectories of soil development and biological community changes, providing insights into the processes driving soil recovery and ecosystem restoration. Another area where CCA is especially useful is in examining the effects of soil contamination or pollution on biological communities. For example, in a study assessing the impact of heavy metals on soil microbial diversity, CCA can be used to relate the concentrations of metals like lead, cadmium, and zinc to shifts in microbial community composition. This analysis helps to identify specific contaminants that are particularly harmful to soil life and can guide remediation efforts by pinpointing the environmental conditions that need to be addressed to restore healthy soil ecosystems (Ng et al. 2024).

CCA is also valuable in agricultural research, where understanding the relationships between soil properties and crop performance is crucial. For example, researchers might use CCA to study how variations in soil fertility, moisture, and organic matter content affect crop yields and the composition of soil microbial communities that support plant growth. By linking these environmental factors to crop success, CCA can help in developing strategies for optimizing soil management practices, such as targeted fertilization or irrigation, to enhance agricultural productivity. In the field of soil ecology, CCA is particularly useful for studying the distribution of soil invertebrates, such as earthworms, nematodes, and arthropods, in relation to environmental gradients. For instance, a study might use CCA to explore how soil texture, organic matter, and moisture influence the distribution of earthworm species across different habitats. The results can provide insights into the ecological niches occupied by different species and how soil management practices, like tillage or organic amendments, might affect their populations.

CCA is also frequently used in studies of soil-plant interactions, where researchers are interested in how soil conditions influence plant community composition. For example, in a study of plant communities in a restored wetland, CCA might be used to analyze how soil pH, moisture, and nutrient levels affect the presence and abundance of different plant species. This analysis can help identify the soil conditions that support desired vegetation types and guide restoration efforts to create suitable habitats for target species. Another application of CCA in soil science is in understanding the effects of climate change on soil ecosystems. For instance, researchers might use CCA to study how changes in temperature and precipitation patterns affect soil moisture, nutrient cycling, and microbial activity, and how these changes, in turn, influence plant and microbial community composition. This approach can help predict the potential impacts of climate change on soil health and guide adaptation strategies to mitigate these effects.

It is also useful in exploring the relationships between soil properties and ecosystem services, such as carbon sequestration, nutrient cycling, and water retention. For example, a study might use CCA to link soil organic carbon levels and nutrient availability to the diversity and functional traits of soil microbial communities. Understanding these relationships can help in managing soils to enhance their capacity to provide critical ecosystem services, contributing to both environmental sustainability and agricultural productivity. CCA is particularly useful in soil science research because it allows for the comprehensive analysis of complex interactions between soil properties and biological communities. Whether studying the effects of environmental gradients, contamination, agricultural practices, or climate change, CCA provides insights that are critical for understanding and managing soil ecosystems. Its ability to handle multiple variables and reveal the underlying patterns in ecological data makes it an indispensable tool for advancing soil science and soil ecology (Kong et al. 2024).

Practical Application in Using Climatic Data in a Scenario of Climatic Changing Affecting Net Primary Production and Soil C Stock

Canonical Correspondence Analysis (CCA) is a valuable tool for exploring the effects of climate change on net primary production (NPP) and soil carbon (C) stocks. By integrating environmental variables with biological responses, CCA helps to reveal how shifts in climate affect both plant growth and soil carbon dynamics. This approach is particularly useful in scenarios where climate change induces complex interactions between climate variables and ecosystem processes. Consider a scenario where researchers are studying a region experiencing increased temperatures and altered precipitation patterns. The goal is to understand how these climate changes impact NPP and soil C stocks. Using CCA, researchers can simultaneously analyze how changes in temperature, precipitation, and other climate variables influence plant productivity and soil carbon content. By relating these variables to NPP and soil C data, CCA helps to identify the key climate factors driving changes in these ecological processes (Wang et al. 2024a).

In the analysis, researchers collect data on soil C stocks and NPP from various sites within the study area, alongside measurements of climate variables such as temperature and precipitation. CCA is then used to relate these climate variables to variations in NPP and soil C across the sites. For example, the analysis might reveal that higher temperatures are associated with increased NPP but reduced soil C stocks due to higher decomposition rates. The strength of CCA lies in its ability to handle multiple environmental variables and their interactions. In this scenario, CCA can disentangle the effects of temperature and precipitation on NPP and soil C stocks, even when these variables are correlated. For instance, if both temperature and precipitation have a significant impact on NPP, CCA can determine their relative contributions and interactions in shaping plant productivity and soil carbon dynamics (Tang et al. 2024).

An important aspect of using CCA in this context is the generation of ordination plots that visually represent the relationships between climate variables, NPP, and soil C. These plots help researchers to identify patterns and trends, such as which climate variables are most strongly associated with changes in NPP and soil C stocks. For example, a plot might show that sites with increased precipitation have higher NPP but also higher soil C stocks, indicating a positive feedback loop between plant growth and carbon sequestration. Interpreting CCA results requires careful consideration of the ecological context. In the scenario of climate change, researchers must account for how shifts in climate variables might affect plant species composition and soil microbial activity, both of which influence NPP and soil C. For instance, changes in plant community composition due to climate shifts might alter the amount of organic matter returned to the soil, affecting soil C stocks (Kesbi et al. 2024).

Validation of CCA results is crucial for ensuring their reliability. Researchers might validate their findings by comparing CCA results across different regions or over time. This helps to confirm whether the identified relationships between climate variables, NPP, and soil C are consistent and robust. For example, validating results in different climate zones or during different seasons can provide insights into how climate impacts on NPP and soil C may vary across different conditions. The practical implications of CCA results in this scenario are significant for managing ecosystems under climate change. For instance, if CCA reveals that increased temperatures are reducing soil C stocks, land management strategies might focus on mitigating temperature effects or enhancing soil carbon sequestration through practices such as increasing vegetation cover or soil conservation.

Additionally, CCA can help guide adaptation strategies by identifying which climate variables have the most significant impact on NPP and soil C. For example, if precipitation is found to be a major driver of NPP and soil C, strategies to manage water resources or enhance soil moisture retention might be prioritized to maintain ecosystem productivity and carbon storage. It provides a comprehensive approach for understanding the effects of climate change on NPP and soil C stocks by integrating multiple climate variables and their interactions. Its ability to handle complex datasets and reveal underlying patterns makes it a powerful tool for predicting and managing the impacts of climate change on ecosystem processes. By applying CCA, researchers and land managers can gain valuable insights into how to adapt and manage ecosystems in a changing climate to sustain productivity and carbon sequestration.

Applying CCA in Soil Science

Applying Canonical Correspondence Analysis (CCA) in soil science offers valuable insights into the complex interactions between soil properties and ecological processes. This method allows researchers to analyze how environmental variables influence soil organisms and their roles in ecosystem functions. To illustrate this, we will explore advanced examples involving beetles in litter decomposition, ants as ecosystem engineers, and termites as bioindicators of soil disturbances (Chernysheva and Fornasier 2024).

In the context of beetles and litter decomposition, CCA can be used to understand how different soil properties affect beetle communities that play a critical role in breaking down organic matter. For example, researchers might collect data on beetle species composition, litter decomposition rates, and various soil variables such as moisture, pH, and nutrient levels. CCA can then relate these soil variables to beetle communities, revealing which factors most strongly influence beetle diversity and their effectiveness in litter decomposition. This analysis helps to identify key soil conditions that support efficient decomposition processes and maintain soil health.

Ants, known for their role as ecosystem engineers, can also be studied using CCA to explore how soil properties and ant activities influence ecosystem functions. Ants modify soil structure through their nesting and foraging activities, which in turn affects soil aeration, water infiltration, and nutrient cycling. By applying CCA, researchers can examine how different soil properties, such as texture and organic matter content, influence ant species composition and their impact on soil structure. For instance, CCA might reveal that higher soil organic matter content supports a more diverse ant community that enhances soil aeration and nutrient distribution, demonstrating the link between soil conditions and ant-mediated ecosystem processes.

Termites, as bioindicators of soil disturbances, provide another example of how CCA can be applied in soil science. Termite communities are sensitive to changes in soil conditions and can reflect alterations in land use or environmental stress. Researchers can use CCA to analyze termite species composition in relation to various soil disturbance indicators, such as compaction, erosion, and contamination. By incorporating data on soil physical and chemical properties, CCA helps to identify how soil disturbances influence termite populations and their distribution. This approach can reveal patterns that indicate soil degradation and provide insights into how disturbances impact soil health and ecosystem functions (Ghafoor et al. 2024).

In addition to these specific examples, CCA can be used to integrate multiple soil variables and biological responses to provide a comprehensive view of soil ecosystem dynamics. For instance, researchers might study how soil moisture, pH, and nutrient levels interact to affect both beetle communities and litter decomposition rates. By analyzing these interactions, CCA can reveal how combined soil properties influence multiple aspects of soil ecology, helping to understand the broader impacts of environmental changes on soil systems. The power of CCA lies in its ability to handle complex, multidimensional datasets and uncover relationships between environmental variables and biological communities. In soil science, this means that researchers can simultaneously explore how different soil factors affect various organisms and ecosystem processes. For example, CCA can elucidate how soil properties influencing beetle populations also affect ant activities and termite communities, providing a holistic understanding of soil ecology (Kaewfoo et al. 2024).

Interpreting CCA results involves careful consideration of the ecological context and the relationships between soil variables and biological responses. Researchers must assess how identified patterns reflect real ecological processes and how they can inform soil management and conservation practices. For instance, understanding how soil pH affects both beetle diversity and litter decomposition can guide soil management strategies to support healthy decomposition processes and maintain soil fertility. Validating CCA results is essential to ensure their robustness and accuracy. Researchers can validate their findings by comparing CCA outcomes across different study sites or over time. This validation helps confirm that identified relationships between soil properties and biological communities are consistent and reliable, providing confidence in the conclusions drawn from the analysis (Neupert et al. 2024).

The practical applications of CCA in soil science are significant for informing soil management and conservation strategies. For example, if CCA reveals that certain soil conditions support diverse and effective beetle communities, management practices can be tailored to maintain or enhance these conditions. Similarly, understanding how ants and termites respond to soil disturbances can guide interventions to mitigate the impacts of land use changes and restore soil health. Thus, applying CCA in soil science offers valuable insights into the complex interactions between soil properties and ecological processes. By analyzing how environmental variables influence soil organisms such as beetles, ants, and termites, researchers can gain a deeper understanding of soil ecology and inform effective management practices. CCA's ability to handle multidimensional data and reveal underlying patterns makes it an indispensable tool for advancing our knowledge of soil systems and their responses to environmental changes.

Software for CCA

Several software packages are commonly used for performing Canonical Correspondence Analysis (CCA), each offering various features tailored to different levels of expertise and research needs. Here's a summary of some of the most popular options:

1. R (vegan package): R is a widely used statistical software environment, and the vegan package is one of the most popular tools for ecological analysis, including CCA.

Features:

- Comprehensive functions for CCA, including model fitting, permutation tests, and plotting.
- Allows for the integration of CCA results with other multivariate analyses.
- Provides extensive documentation and community support.

Advantages:

- Highly customizable and flexible for various types of ecological data.
- Open-source and freely available.
- Can handle large datasets and complex models.

Usage: Suitable for researchers familiar with coding in R and who require detailed, customizable analyses.

2. CANOCO for Windows: CANOCO is a specialized software package designed for multivariate analysis in ecology, including CCA.

Features:

- User-friendly graphical interface with built-in tools for CCA.
- Provides options for handling various types of ecological data and experimental designs.
- Offers advanced visualization options, including ordination plots and biplots.

Advantages:

- Intuitive for users who prefer a graphical interface over coding.
- Includes comprehensive documentation and tutorials.
- Suitable for both beginners and advanced users.

Usage: Ideal for ecologists who need a robust tool for CCA with a focus on ease of use and visual output.

3. PRIMER-E: It is a software package for multivariate statistical analysis, including CCA, widely used in ecological research.

Features:

- Comprehensive tools for multivariate analysis, including CCA and related methods.
- Allows for sophisticated data exploration and visualization.
- Provides options for permutation tests and model validation.

Advantages:

- User-friendly with a strong focus on ecological data.
- Good for data exploration and visualization with interactive plots.
- Includes support and training resources.

Usage: Suitable for researchers looking for an integrated package that combines multivariate analysis with data visualization.

4. PAST (Paleontological Statistics): It is a free software package designed for statistical analysis in paleontology and ecology, including CCA.

Features:

- Offers tools for CCA along with a variety of other statistical and multivariate analyses.
- Provides graphical output for visualizing ordination results.

Advantages:

- Free and accessible, making it a good choice for budget-conscious researchers.
- Easy to use with a straightforward interface.

Usage: Ideal for those needing a free, easy-to-use tool for basic CCA and other statistical analyses.

5. EcoSimR: It is a software package specifically designed for ecological modeling and analysis, including CCA.

Features:

- Provides tools for analyzing ecological data and running CCA.
- Includes functions for simulation and analysis of ecological patterns.

Advantages:

- Tailored for ecological studies with a focus on simulations.
- Free and open-source with good documentation.

Usage: Suitable for researchers interested in combining CCA with ecological simulations and modeling.

6. XLSTAT: It is an add-on for Microsoft Excel that offers various statistical analyses, including CCA.

Features:

- Integrates with Excel, making it accessible for users familiar with this platform.
- Provides tools for multivariate analysis and data visualization.

Advantages:

- Easy to use for those who prefer working within Excel.
- Includes a range of statistical tools beyond CCA.

Usage: Ideal for users who need a straightforward tool for CCA within the familiar Excel environment.

Each software package has its strengths and is suited to different levels of expertise and types of research needs. R (vegan) and CANOCO for Windows are among the most widely used due to their comprehensive features and strong support for ecological data analysis. PRIMER-E and PAST are also popular for their user-friendly interfaces and integration with data visualization. EcoSimR offers a specialized approach for ecological modeling, while XLSTAT provides a convenient option for users who prefer working within Excel.

Example Application in Soil Science

Canonical Correspondence Analysis (CCA) is a sophisticated tool in soil science that allows researchers to explore complex relationships between soil properties, ecological variables, and environmental factors. One advanced application of CCA is in understanding nutrient cycling in agricultural soils. By applying CCA, researchers can examine how soil chemical properties, such as nitrogen (N) and phosphorus (P) levels, interact with soil microbial communities to influence nutrient cycling processes. For instance, a study might use CCA to analyze how variations in soil N and P availability correlate with changes in microbial biomass and activity across different crop management practices. This approach helps identify key soil and microbial factors that drive nutrient cycling, allowing for more effective management strategies to optimize soil fertility and crop productivity (Zhang et al. 2024c).

Another advanced application of CCA is in the study of soil ecology and the impact of soil physical and chemical properties on microbial communities. By incorporating variables such as soil texture, moisture, and organic carbon content, CCA can reveal how these factors influence the distribution and diversity of soil microorganisms. For example, researchers might use CCA to assess how variations in soil moisture and organic matter affect the composition of fungal and bacterial communities in forest soils. This analysis provides insights into the ecological interactions within the soil environment and helps in understanding how soil management practices or environmental changes impact soil health and microbial diversity.

CCA is also valuable in examining the effects of biological invasion on soil ecosystems. When invasive plant species alter soil properties and nutrient availability, CCA can be used to investigate the changes in soil microbial communities and nutrient dynamics. For instance, a study might apply CCA to analyze how the invasion of a non-native plant species affects soil nutrient levels, microbial community structure, and soil enzyme activities compared to native plant communities. This application helps in understanding the ecological consequences of biological invasions and informs management practices to mitigate their impact on soil health and ecosystem functioning.

Carbon (C) dynamics in soils is another area where CCA can provide advanced insights. By analyzing how soil properties and management practices influence soil organic carbon (SOC) pools and their turnover, CCA can help identify factors that contribute to carbon sequestration or loss. For example, researchers might use CCA to study how different land use practices, such as conservation tillage versus conventional tillage, affect SOC dynamics in agricultural soils. This analysis helps in assessing the effectiveness of various land management strategies for enhancing soil carbon storage and addressing climate change.

Climate change is a pressing issue that significantly impacts soil properties and processes. CCA can be employed to explore how changes in climate variables, such as temperature and precipitation, affect soil properties and ecosystem functions. For instance, a study might use CCA to examine how increasing temperatures and altered precipitation patterns influence soil moisture, nutrient availability, and microbial activity in semi-arid regions. This application helps in understanding the potential impacts of climate change on soil health and informs adaptation strategies to maintain soil productivity and sustainability.

Soil disturbance, such as those caused by agricultural practices, construction, or land use changes, can be analyzed using CCA to assess its impact on soil properties and ecological functions. Researchers might use CCA to evaluate how different types and intensities of soil disturbance affect soil physical properties, nutrient availability, and microbial communities. For example, a study could analyze the effects of different tillage practices on soil structure, compaction, and microbial diversity, providing insights into how to mitigate the negative effects of soil disturbance and promote soil health.

In addition to these applications, CCA is useful in exploring the relationships between soil properties and vegetation dynamics. By incorporating both soil and vegetation data, researchers can use CCA to analyze how soil conditions influence plant community composition and distribution. For example, a study might apply CCA to investigate how variations in soil pH, nutrient levels, and moisture affect the diversity and abundance of plant species in a grassland ecosystem. This approach helps in understanding the interplay between soil and vegetation and informs habitat management and restoration efforts. CCA also plays a role in studying the impact of land management practices on soil health. By analyzing how different management practices, such as crop rotation, organic amendments, and conservation tillage, affect soil properties and ecological functions, researchers can use CCA to identify practices that enhance soil health and sustainability. For instance, a study might use CCA to assess how various organic amendment applications influence soil nutrient levels, microbial activity, and plant growth in a horticultural setting. This application helps in developing evidence-based recommendations for soil management practices.

Actually, CCA can be employed to evaluate the effects of soil contamination on soil health and ecosystem functions. By incorporating data on soil contaminants, such as heavy metals or pesticides, and soil properties, CCA can reveal how contamination influences soil microbial communities, nutrient availability, and plant health. For example, a study might use CCA to investigate how soil contamination from agricultural runoff affects microbial diversity and soil enzyme activities in affected areas. This analysis provides valuable information for managing and mitigating the impacts of soil contamination on soil and environmental health.

Conclusion

Canonical Correspondence Analysis (CCA) stands out as a powerful and versatile tool in soil science and data analysis, offering a robust framework for exploring complex relationships between environmental variables and biological communities. Its ability to integrate multiple datasets, including soil properties and species compositions, makes it invaluable for understanding how various factors interact to influence soil health and ecosystem functions. By elucidating the underlying patterns and associations within high-dimensional data, CCA enables researchers to uncover critical insights into the impacts of environmental changes, management practices, and ecological processes. This capacity to reveal and interpret intricate relationships is crucial for advancing soil science, guiding effective land management strategies, and addressing pressing environmental challenges such as climate change and biodiversity loss. As such, CCA not only enhances our understanding of soil ecosystems but also supports the development of targeted, evidence-based approaches for preserving and improving soil health and ecological integrity.

References

Bashtian MH, Karimi A, Sepehr A, Lakzian A, Caballero ER (2024) Spatial relationship of landform surface features and biocrusts, and their effect on soil microbial biomass on an alluvial fan. Earth Surf Process Landf 49:1348–1360. https://doi.org/10.1002/esp.5771

Chernysheva EV, Fornasier F (2024) Specific features of the pool of hydrolytic enzymes in soils of agricultural terraces in the eastern Caucasus. Eurasian Soil Sci 57:995–1006. https://doi.org/10.1134/S1064229324600076

Dayang Najwa AB, Elexson N, Dalene L, Teng ST (2024) *Vibrio* species and cyanobacteria: understanding their Association in Local Shrimp Farm Using Canonical Correspondence Analysis (CCA). Microb Ecol 87:e51. https://doi.org/10.1007/s00248-024-02356-5

Ghafoor AZ, Derejko A, Studnicki M (2024) Identification of plant and soil characteristics affecting stability of winter wheat cultivar in temperate climates. Agronomy 14:e779. https://doi.org/10.3390/agronomy14040779

Kaewfoo M, Bunyavejchewin S, Marod D et al (2024) Termitaria enhance soil and forest diversity in deciduous dipterocarp forest, Northern Thailand. J Trop Ecol 40:e5. https://doi.org/10.1017/S0266467423000342

Kesbi MA, Erfanzadeh R, Fattahi B (2024) Similarity between soil seed bank and standing vegetation and their relationship with soil and topographical characteristics in a riparian zone. Community Ecol 25:89–101. https://doi.org/10.1007/s42974-023-00180-4

Kong C, Zhang S, Yuan S et al (2024) Soil bacterial community characteristics and its effect on organic carbon under different fertilization treatments. Front Microbiol 15:e1356171. https://doi.org/10.3389/fmicb.2024.1356171

Neupert M, Aubert M, Langlois E (2024) Tidal restriction does not always supplant local conditions shaping estuarine plant communities. Estuar Coast Shelf Sci 305:108857. https://doi.org/10.1016/j.ecss.2024.108857

Ng W, Winowiecki LA, Karari V et al (2024) Exploring mid-infrared spectral transfer functions for the prediction of multiple soil properties using a global dataset. Soil Sci Soc Am J 88:1234–1247. https://doi.org/10.1002/saj2.20697

Rivera-Zizumbo AA, Velez P, Ojeda M et al (2024) Implementing ITS1 metabarcoding for the analysis of soil microeukariotic diversity in the mountain cloud forest. J Soils Sediments 24:3001–3012. https://doi.org/10.1007/s11368-024-03839-1

Tang C, Hou J, Liu D et al (2024) Applying fluorescence spectroscopy coupled with Gaussian band fitting to reveal dynamic variation process of humus fractions from riparian soils along an urbanized river. Sci Total Environ 927:e172193. https://doi.org/10.1016/j.scitotenv.2024.172193

Tenaw TB, Gode TB, Molla EL, Woldemariam ZA (2024) Topography and soil variables drive the plant community distribution pattern and species richness in the Arjo-Diga forest in western Ethiopia. PLoS One 19:e0307888. https://doi.org/10.1371/journal.pone.0307888

Wang W, Wang CM, Ho YC et al (2024a) Plant communities of the tern sanctuary on the Matsu Islands as a breeding habitat for seabirds. Diversity 16:e501. https://doi.org/10.3390/d16080501

Wang Y, Xu J, Yuan Q et al (2024b) Composition and diversity of soil microbial communities change by introducing *Phallus impudicus* into a *Gastrodia elata* Bl.-based soil. BMC Microbiol 24:204. https://doi.org/10.1186/s12866-024-03330-4

Xing S, Wang WJ, Wang L et al (2024) Soil nutrient content dominates short-term vegetation changes in alpine tundra of Changbai Mountains. Front Microbiol 15:1422529. https://doi.org/10.3389/fmicb.2024.1422529

Zhang F, Xie Y, Peng R, Ji X, Bai L (2024a) Heavy metals and nutrients mediate the distribution of soil microbial community in a typical contaminated farmland of South China. Sci Total Environ 947:e174322. https://doi.org/10.1016/j.scitotenv.2024.174322

Zhang L, Ning C, Qi S et al (2024b) Understory specific species appearances indicate the soil improvement of low-function *Platycladus orientalis* forest in Beijing's mountainous areas. Ecol Indic 162:111996. https://doi.org/10.1016/j.ecolind.2024.111996

Zhang S, Chen T, Chen Y et al (2024c) A comparative analysis of bacterial and fungal communities in coastal and inland pecan plantations. Microorganisms 12:e1313. https://doi.org/10.3390/microorganisms12071313

Zhou S, Lv Y, Song Z, Bi X, Meng L (2024) Long-term fertilization contributes to carbon saturation in neutral-to-alkaline soils but not in acidic soils. Commun Soil Sci Plant Anal 55:1224–1235. https://doi.org/10.1080/00103624.2024.2303095

Zhu Z, Chen J, Hu H et al (2024) Soil quality evaluation of different land use modes in small watersheds in the hilly region of southern Jiangsu. Ecol Indic 160:e111895. https://doi.org/10.1016/j.ecolind.2024.111895

Structural Equation Modelling (SEM)

8

Abstract

The primary difference between Structural Equation Modeling (SEM) and multiple regression lies in SEM's ability to model complex relationships, including indirect effects and latent variables, while multiple regression focuses on direct relationships between observed variables. The main assumptions required for SEM include the correct specification of the model, multivariate normality, and sufficient sample size. SEM handles high-dimensional soil data effectively by integrating multiple observable variables, such as essential macronutrients, to estimate latent variables like soil fertility and plant nutrition. This approach makes SEM particularly useful in soil science, where it can unravel the intricate interactions among soil physical, chemical, and biological properties. Before applying SEM, researchers should ensure model specification accuracy, check for data normality, assess sample size adequacy, and verify the appropriateness of their data structure to ensure reliable results.

Keywords

Latent variables · Model fit · Path analysis · Soil Science · Structural equation modelling (SEM)

Questions Covered in the Chapter

(a) What is the primary difference between SEM and multiple regression?
(b) What are the main assumptions required for SEM?
(c) How does SEM handle high-dimensional soil data?
(d) Why is SEM particularly useful in soil science research?
(e) What should researchers check for before applying SEM to ensure reliable results?

T. Souza, *Advanced Statistical Analysis for Soil Scientists*,
https://doi.org/10.1007/978-3-031-88161-9_8

Introduction

Structural Equation Modelling (SEM) is an advanced statistical technique that allows researchers to test complex relationships among observed and latent variables. It integrates multiple regression, factor analysis, and path analysis, enabling the evaluation of direct and indirect effects within a system. SEM is particularly useful in fields like soil science, where relationships between variables can be intricate and multifaceted. SEM relies on several key assumptions. First, the relationships between variables must be linear. Second, the data should follow a multivariate normal distribution, particularly when using Maximum Likelihood Estimation. Third, SEM assumes that the sample size is sufficiently large to provide reliable estimates, typically requiring a minimum of 200 cases. These assumptions ensure that the model's estimates are accurate and that the conclusions drawn are robust (Zhu et al. 2024).

One of the main advantages of SEM is its ability to model complex systems with multiple dependent and independent variables. Unlike traditional regression models that typically examine one dependent variable at a time, SEM can analyze multiple relationships simultaneously. This feature is particularly advantageous in soil science, where variables like soil properties, plant growth, and microbial activity often interact in complex ways. SEM is unique in its ability to incorporate latent variables—unobserved constructs that are inferred from observed variables. This is particularly useful in soil science when dealing with constructs like soil quality or fertility, which may not be directly measurable but can be represented by multiple indicators such as organic matter content, pH, and nutrient levels (Xu et al. 2024).

SEM is a powerful tool for testing theoretical models and hypotheses. In soil science, researchers can use SEM to evaluate models that describe the relationships between soil properties and ecological processes. For example, SEM can be used to test hypotheses about how soil texture and organic matter content influence water retention and plant growth. One of the strengths of SEM is its ability to estimate and test causal relationships through path analysis. This is particularly relevant in soil science, where understanding the causal links between soil management practices and soil health is crucial for developing sustainable agricultural systems (Yang et al. 2024a).

SEM provides various indices to assess the fit of a model, such as the Chi-square test, Root Mean Square Error of Approximation (RMSEA), and Comparative Fit Index (CFI). These indices help researchers evaluate whether their model adequately represents the data. In soil science, ensuring a good model fit is essential for drawing valid conclusions about the relationships between soil variables. Another advantage of SEM is its ability to account for measurement error. In soil science, measurement error can arise from various sources, such as sampling variability or instrument precision. SEM can adjust for these errors, providing more accurate estimates of the relationships between variables (Le Geay et al. 2024).

SEM's ability to model complex relationships makes it a valuable tool in soil science. For instance, researchers can use SEM to study the interactions between soil structure, nutrient cycling, and microbial activity. This holistic approach is essential for understanding soil ecosystems and developing strategies to improve soil health and fertility. SEM has been applied in various soil science studies, such as examining the effects of land use on soil properties, understanding the drivers of soil biodiversity, and exploring the interactions between soil and plant communities. These applications demonstrate SEM's versatility and potential to advance soil science research. As soil science continues to evolve, the use of SEM is likely to increase, particularly in studies that require a comprehensive understanding of complex ecological interactions. By integrating SEM with other advanced techniques, such as Geographic Information Systems (GIS) and remote sensing, soil scientists can gain deeper insights into soil processes and contribute to more sustainable land management practices (Chen et al. 2024).

Main Differences Between SEM and Multiple Regression

The primary difference between Structural Equation Modelling (SEM) and multiple regression lies in their capacity to handle complex relationships among variables and their underlying assumptions about the data. Multiple regression is a straightforward technique that models the relationship between a single dependent variable and one or more independent variables. It provides a way to predict the value of the dependent variable based on the independent variables, assuming that the relationships are linear and that there is no interdependence among the predictors. However, this simplicity also limits multiple regression, particularly in fields like soil science, where variables often interact in complex and interdependent ways (Carrasco-Espinosa et al. 2024).

SEM, on the other hand, extends beyond the capabilities of multiple regression by allowing the modeling of multiple dependent and independent variables simultaneously. It can incorporate both observed variables, such as soil texture or soil organic matter content, and latent variables, which represent constructs that are not directly measurable, like soil quality or nutrient cycling potential. SEM also enables researchers to specify and test complex hypotheses about the relationships among these variables, including indirect effects and feedback loops. For example, in a study of soil properties, SEM could model the direct effect of soil texture on water retention and the indirect effect mediated by soil organic matter (Hu et al. 2024a).

Consider a scenario where a researcher is interested in understanding how soil texture influences soil water retention and nutrient cycling, with soil organic matter serving as a mediator. Using multiple regression, the researcher could analyze the direct relationship between soil texture and water retention, and separately between soil texture and nutrient cycling. However, multiple regression would not allow the researcher to simultaneously account for the indirect effects of soil texture on these outcomes through changes in soil organic matter content. This is where SEM excels, as it can model the entire system of relationships in one comprehensive framework, providing insights into both the direct and indirect effects of soil texture on water retention and nutrient cycling (da Silva et al. 2024).

Another important distinction is how each method handles measurement error. In multiple regression, measurement error in the independent variables is generally assumed to be negligible or is simply ignored, which can lead to biased estimates and misleading conclusions. In contrast, SEM explicitly accounts for measurement error, providing more accurate estimates of the relationships among variables. This is particularly important in soil science, where measurement error can arise from variability in sampling methods or precision of laboratory instruments. For instance, when measuring soil organic matter, slight errors in quantification can significantly impact the analysis of its effect on water retention and nutrient cycling (Li et al. 2024).

SEM also allows for the testing of complex causal models, something multiple regression does not support. In multiple regression, the direction of the relationship between variables is typically assumed, but it does not provide a way to test these assumptions. SEM, however, allows researchers to propose and test specific causal pathways. For example, a soil scientist might hypothesize that soil texture influences nutrient cycling primarily through its effect on soil organic matter, which in turn affects water retention. SEM can be used to test this hypothesis by modeling the causal relationships and examining the strength and significance of each pathway. Furthermore, SEM's ability to model latent variables offers a significant advantage over multiple regression. Latent variables, which are inferred from multiple observed indicators, can represent complex constructs like soil fertility or ecosystem health. In multiple regression, the analysis is limited to the observed variables, which might not fully capture the underlying processes of interest. SEM allows these latent variables to be included in the model, providing a more comprehensive understanding of the system. For example, a latent variable representing soil health could be modeled based on indicators such as organic matter content, nutrient availability, and microbial activity, offering deeper insights into the factors driving nutrient cycling (Hu et al. 2024b).

The flexibility of SEM in modeling complex systems also extends to its capacity to analyze both recursive and non-recursive relationships. In multiple regression, relationships are typically assumed to be one-way; that is, changes in the independent variables cause changes in the dependent variable. SEM, however, can model bidirectional relationships, where variables influence each other. This is particularly relevant in soil science, where feedback loops are common. For instance, changes in soil organic matter can affect soil water retention, which in turn influences organic matter decomposition rates, creating a feedback loop that multiple regression cannot adequately capture. In other words, while multiple regression is a useful tool for analyzing simple relationships between variables, it falls short when dealing with the complexity inherent in many soil science systems. SEM provides a more sophisticated framework that accounts for multiple, interrelated variables, measurement error, latent constructs, and complex causal relationships. By using SEM, soil scientists can gain a deeper understanding of how soil texture, organic matter, water retention, and nutrient cycling interact, leading to more informed decisions about soil management and conservation (Moshagen and Bader 2024).

Key Concepts of SEM

Structural Equation Modelling (SEM) offers a powerful way to explore complex relationships in soil science, where variables often interact in intricate ways. To grasp SEM, it's crucial to understand that it goes beyond traditional statistical methods like multiple regression by allowing the simultaneous analysis of multiple variables and their interconnections. For instance, when studying soil chemical properties and their impact on plant nutrition, SEM enables the analysis of direct and indirect effects, such as how soil pH might directly influence nutrient availability and indirectly affect plant growth through changes in microbial activity (Na et al. 2024).

A key concept in SEM is the distinction between observed and latent variables. Observed variables are directly measured, such as soil pH or nitrogen content, while latent variables represent underlying constructs that are not directly measured but are inferred from observed data. In the context of soil quality, a latent variable could be constructed to represent overall soil health, incorporating indicators like organic matter, nutrient levels, and microbial biomass. SEM allows researchers to model these latent variables and assess how they influence other aspects of the ecosystem, such as net primary production. Another important aspect of SEM is its ability to handle complex causal relationships. Unlike simpler methods that only explore direct effects, SEM can model both direct and indirect pathways. For example, soil chemical properties like cation exchange capacity (CEC) may directly influence nutrient retention, but their effect on plant nutrition might be mediated by other factors such as soil moisture. SEM can untangle these pathways, offering a more nuanced understanding of how different soil properties work together to affect plant growth and overall ecosystem productivity (Ren et al. 2024).

Measurement error is a challenge in any scientific study, and SEM is particularly well-suited to address it. In soil science, measurement error might arise from variability in sampling methods or instrument precision. SEM explicitly incorporates measurement error into the model, allowing for more accurate estimates of the relationships among variables. This is especially important when studying plant nutrition, where small errors in measuring nutrient levels could lead to misleading conclusions about their impact on plant health. SEM also excels at testing theoretical models that describe how different soil and plant processes are interconnected. For example, a researcher might develop a model that hypothesizes the relationships between soil quality, nutrient cycling, and net primary production. SEM allows for the testing of such models, providing statistical evidence to support or refute the proposed relationships. This capability is invaluable in advancing our understanding of soil-plant interactions and improving agricultural practices (Jarvis et al. 2024).

A unique feature of SEM is its ability to incorporate feedback loops, where variables influence each other reciprocally. In soil science, this is particularly relevant when considering processes like nutrient cycling, where plant uptake of nutrients can alter soil chemical properties, which in turn affect further nutrient availability. SEM allows these bidirectional relationships to be modeled, offering insights into the dynamic nature of soil and plant systems that simpler methods cannot capture.

The concept of model fit is central to SEM. After constructing a model, researchers must assess how well it represents the data. Various fit indices, such as the Comparative Fit Index (CFI) or Root Mean Square Error of Approximation (RMSEA), help determine whether the model is a good representation of the observed relationships. For instance, if a model of soil quality and net primary production shows poor fit, it may indicate that key variables or pathways have been overlooked, prompting a revision of the model (Liu et al. 2024).

SEM's flexibility also extends to its ability to compare multiple models. In soil science, this could involve comparing different hypotheses about how soil chemical properties influence plant nutrition. For example, one model might suggest that soil organic matter directly influences plant growth, while another proposes that its effect is mediated through soil microbial activity. SEM allows researchers to compare these models and determine which best explains the observed data, leading to more robust conclusions. In applying SEM to soil science, it's essential to recognize that the method is not just about crunching numbers; it's about building and testing theories that explain how soil and plant systems function. By providing a framework for integrating multiple variables and testing complex hypotheses, SEM helps scientists move beyond simple correlations and develop a deeper understanding of the underlying processes that drive soil health, plant nutrition, and ecosystem productivity (Wang et al. 2024).

For young scientists in soil science, mastering SEM opens up new possibilities for research. Whether studying the effects of soil chemical properties on plant nutrition or exploring the factors that drive net primary production, SEM provides the tools needed to tackle complex questions and contribute to the growing field of soil science. As our understanding of these systems deepens, the insights gained through SEM will play a crucial role in developing sustainable agricultural practices and preserving soil health for future generations (Xie et al. 2024).

Main Assumptions Required for SEM

When conducting Structural Equation Modelling (SEM), it is crucial to understand the assumptions that underlie the method to ensure valid and reliable results. The first major assumption is that the relationships between variables are linear. SEM is designed to model linear associations, meaning that changes in one variable are expected to produce proportional changes in another. In the context of soil science, this assumption implies that, for instance, an increase in soil organic matter will lead to a predictable and proportional improvement in soil quality or plant nutrient availability. If the relationship between variables is non-linear, such as a threshold effect where soil quality only improves significantly after a certain level of organic matter is reached, SEM might not capture this accurately, leading to biased results (Xu et al. 2024).

Another important assumption in SEM is the requirement for multivariate normality. This means that the variables in the model should follow a normal distribution when considered individually and together. In soil science, this assumption is particularly relevant when dealing with variables like soil nutrient concentrations or yield measurements in food production. If the data deviates significantly from normality, the estimates and tests within the SEM framework might become unreliable. For example, if nutrient concentration data is highly skewed due to outliers or extreme values, this could distort the model's findings on how nutrient dynamics affect plant growth and food production. A sufficient sample size is also a critical assumption in SEM. Because SEM involves estimating multiple parameters simultaneously, it requires a large sample to provide stable and accurate estimates. The general rule of thumb is to have at least 200 cases, but the required sample size may vary depending on the complexity of the model. In soil science research, this might involve collecting extensive data on soil quality indicators, plant nutrient levels, and crop yields across multiple locations and time points. Without a sufficiently large sample, the model may produce unstable estimates, leading to less confidence in the findings about the interactions between soil properties and plant nutrient dynamics (Yang et al. 2024b).

The assumption of no perfect multicollinearity among the predictors is another key consideration in SEM. Multicollinearity occurs when two or more variables are highly correlated, which can make it difficult to isolate the individual effect of each variable. In the context of soil science, variables like soil pH and cation exchange capacity (CEC) might be highly correlated, complicating the analysis of how each independently affects plant nutrient availability. If multicollinearity is present, it can lead to inflated standard errors and unreliable parameter estimates, ultimately weakening the conclusions drawn from the model. SEM also assumes that the model is correctly specified, meaning that the relationships between variables are accurately represented. This includes ensuring that all relevant variables are included and that the direction of the relationships is correctly identified. For example, in a model examining the impact of soil quality on food production, failing to include an important variable like irrigation or assuming an incorrect causal direction between nutrient availability and crop yield could lead to a misrepresentation of the true relationships. Model misspecification can result in biased estimates and incorrect conclusions, highlighting the importance of careful model construction (Zhao et al. 2024).

Another assumption is that the model should reflect the underlying causal structure of the relationships between variables. SEM is often used to test theoretical models that describe how variables interact. In soil science, this might involve modeling how soil nutrient dynamics influence plant growth, which in turn affects food production. The assumption here is that the model structure accurately represents the causal pathways. If the model does not align with the actual causal relationships, the results may be misleading. For example, if the model incorrectly assumes that nutrient availability directly impacts food production without considering the mediating role of plant health, it could lead to incorrect inferences.

The assumption of independent observations is also crucial in SEM. This means that the data points should not influence each other. In soil science, this assumption could be violated if, for instance, soil samples from different plots are not truly independent due to shared environmental conditions or management practices. If observations are correlated, such as in cases where nearby plots share similar soil quality characteristics, the independence assumption is violated, which can lead to underestimated standard errors and overconfidence in the model's results (Yang et al. 2024c).

Finally, SEM assumes that the errors in the model are uncorrelated. In other words, the residuals (the differences between observed and predicted values) should not be systematically related to each other. If there are correlations among the errors, this suggests that the model is missing some important relationships or variables. In soil science, correlated errors might occur if there are unaccounted-for factors influencing both soil quality and food production, such as local climate conditions or specific management practices. Correlated errors can lead to biased estimates and reduce the validity of the model's conclusions. Understanding and meeting these assumptions is essential for effectively applying SEM in soil science research. By carefully considering the linearity of relationships, ensuring multivariate normality, obtaining a sufficient sample size, and addressing multicollinearity, researchers can build robust models that accurately reflect the complex interactions between soil quality, plant nutrient dynamics, and food production. This rigorous approach allows for more reliable insights into how these factors contribute to sustainable agricultural practices and soil management strategies (Sterner et al. 2024).

Steps to Conduct SEM

When embarking on a Structural Equation Modelling (SEM) analysis in the field of soil science and plant nutrition, the first step is to develop a clear research question or hypothesis. This question should address the relationships you aim to explore, such as how soil physical properties influence the availability of plant nutrients in saline soils. By defining a precise research question, you set the foundation for constructing a model that reflects the underlying processes you wish to investigate. For example, you might hypothesize that soil texture directly affects water retention, which in turn influences the mobility of nutrients like sodium and potassium in saline environments (Sheikh et al. 2024).

Next, it's important to conceptualize your model by identifying the key variables involved in your study. These variables can be observed, like soil salinity or plant nutrient concentrations, or latent, such as soil health or nutrient cycling efficiency. In this stage, you need to think critically about the potential relationships between these variables. For instance, in saline soils, you might consider how soil compaction affects root penetration and subsequently impacts nutrient uptake by plants. By carefully selecting and defining these variables, you create a conceptual map that guides your SEM analysis.

Once your variables are identified, the next step is to specify the model by outlining the hypothesized relationships among these variables. In SEM, this involves creating a diagram that visually represents the paths connecting different variables. For example, you could draw arrows from soil salinity and texture to plant nutrient availability, with additional paths showing how these factors influence plant growth.

This visual model helps clarify the assumed causal relationships and allows for the identification of direct and indirect effects. The specificity of this model is crucial, as it will determine how the SEM will be structured and interpreted.

After specifying the model, you need to collect the data required to test your hypotheses. In soil science and plant nutrition, this might involve gathering data on soil texture, salinity levels, and plant nutrient concentrations from various field sites or experimental plots. The quality and quantity of this data are paramount, as SEM relies on large, accurate datasets to produce reliable results. For instance, you might collect soil samples from different saline environments and measure their physical properties alongside plant tissue analysis to assess nutrient levels. This data collection process should be thorough and systematic to ensure that your SEM analysis is based on robust and representative information.

With your data in hand, the next step is to assess the adequacy of the data for SEM analysis. This involves checking for issues like missing data, multicollinearity, and ensuring that the variables meet the assumptions required for SEM, such as normality and linearity. For example, you might discover that soil salinity data is skewed due to extreme values in certain plots. In this case, you may need to transform the data or consider alternative approaches to address these issues. Ensuring that your data is suitable for SEM is critical for the accuracy and validity of the model.

Once you have prepared the data, the next stage is to estimate the parameters of your model using SEM software. This involves using statistical algorithms to calculate the strength and significance of the relationships between variables in your model. For instance, the software might estimate the direct effect of soil texture on nutrient availability and the indirect effect through changes in soil structure in saline soils. The estimation process provides the numerical values that will be used to interpret the relationships and draw conclusions about your hypotheses.

After estimating the parameters, it is essential to evaluate the overall fit of the model. This step involves using various fit indices, such as the Chi-square test, Comparative Fit Index (CFI), and Root Mean Square Error of Approximation (RMSEA), to assess how well the model represents the observed data. If the model shows a good fit, it suggests that the hypothesized relationships align with the actual data, providing support for your theoretical framework. For example, a good fit in your model might indicate that your assumptions about how soil physical properties influence nutrient dynamics in saline soils are accurate (Michalski and Zaleski 2024).

If the model fit is not satisfactory, the next step is to refine the model by making adjustments based on the fit indices and modification indices provided by the SEM software. This might involve adding or removing paths, re-specifying relationships, or even reconsidering the inclusion of certain variables. In the context of saline soils, you might realize that an additional factor, like soil organic matter content, should be included in the model to better explain nutrient availability. Refining the model is an iterative process that helps improve its accuracy and relevance.

Once the model has been refined and demonstrates a good fit, the final step is to interpret the results and draw conclusions about the relationships among the variables. This involves examining the estimated parameters to understand the magnitude and direction of the effects. For instance, you might find that soil texture has a strong positive effect on nutrient retention in saline soils, while soil compaction negatively impacts nutrient uptake by plants. These interpretations provide valuable insights into the underlying processes and can inform soil management practices in saline environments.

Communicating the results of your SEM analysis is crucial for advancing knowledge in soil science and plant nutrition. This involves writing a clear and comprehensive report or presentation that explains the research question, the model specification, the data collection process, and the findings. By effectively communicating your results, you contribute to the scientific community's understanding of how soil physical properties influence plant nutrition in challenging environments like saline soils.

Finally, it's important to reflect on the implications of your findings for soil management and agricultural practices. The insights gained from SEM can guide decision-making, such as selecting appropriate soil amendments to improve nutrient availability in saline soils or optimizing irrigation practices to reduce soil compaction. By applying these findings, you can help develop strategies that enhance soil quality and plant nutrition, ultimately contributing to sustainable food production in areas affected by soil salinity (Zhu et al. 2024).

Handling High-Dimensional Soil Data with SEM

Structural Equation Modelling (SEM) is well-suited to handle high-dimensional data, such as that found in soil science, where multiple variables interact in complex ways. High-dimensional data refers to datasets with a large number of variables, which can make it challenging to analyze and interpret using traditional statistical methods. SEM, however, is capable of managing this complexity by allowing for the simultaneous analysis of multiple relationships, both direct and indirect, among observed and latent variables.

In the context of soil science, imagine you are studying the impact of all essential macronutrients—nitrogen (N), phosphorus (P), potassium (K), calcium (Ca), magnesium (Mg), and sulfur (S)—on soil fertility and plant nutrition. These macronutrients are your observable variables, which you can directly measure in soil samples. However, soil fertility and plant nutrition are more abstract concepts that are not directly measurable; they are considered latent variables. SEM can model the influence of these macronutrients on these latent variables, accounting for the complex interactions among them (Yang et al. 2024a, b, c).

For example, soil fertility could be modeled as a latent variable influenced by the levels of N, P, K, Ca, Mg, and S. In this case, SEM allows you to explore how each macronutrient contributes to overall soil fertility, taking into account that these nutrients do not act in isolation. Nitrogen might have a direct positive effect on soil fertility, while potassium might have both a direct effect and an indirect effect mediated through its interaction with magnesium. SEM handles these multiple pathways, providing a comprehensive understanding of how each nutrient influences soil fertility.

Similarly, plant nutrition can be treated as another latent variable influenced by both soil fertility and the individual macronutrient levels. SEM allows you to model how soil fertility (itself a latent variable derived from the observable macronutrients) affects plant nutrition, and how specific nutrients might directly or indirectly influence plant growth. For instance, phosphorus might have a strong direct impact on plant nutrition, while calcium's effect could be more indirect, mediated through its role in soil structure and root development (Xie et al. 2024).

One of the advantages of SEM in handling high-dimensional soil data is its ability to incorporate and manage multicollinearity, which is common when dealing with many correlated soil variables. For instance, in soil science, macronutrients like calcium and magnesium are often correlated, making it difficult to separate their individual effects using traditional regression techniques. SEM, however, can account for these correlations within the model, allowing you to estimate the unique contribution of each macronutrient to soil fertility and plant nutrition without the bias that multicollinearity might introduce.

Moreover, SEM's capacity to model latent variables is particularly useful when dealing with high-dimensional data, where the goal is often to reduce the dimensionality of the data by summarizing the information contained in multiple observed variables. For example, rather than analyzing the effects of each macronutrient individually, SEM allows you to combine them into a latent variable representing overall soil fertility. This reduces the complexity of the analysis while still capturing the essential information needed to understand how soil conditions affect plant nutrition.

SEM also offers robust tools for assessing model fit, which is crucial when dealing with high-dimensional data. The fit indices provided by SEM, such as the Comparative Fit Index (CFI) and Root Mean Square Error of Approximation (RMSEA), help determine whether the model accurately represents the relationships in the data. If the model does not fit well, you can modify it by adding or removing paths or latent variables until a satisfactory model is achieved, ensuring that your analysis remains both accurate and interpretable despite the complexity of the data.

Additionally, SEM can be used to test complex hypotheses that involve multiple layers of relationships among high-dimensional data. For example, you might hypothesize that soil fertility (influenced by macronutrients) not only affects plant nutrition directly but also indirectly through its impact on soil microbial activity, another latent variable. SEM can model these hypotheses, allowing you to test whether the proposed pathways are supported by the data, providing deeper insights into how macronutrients interact to influence both soil fertility and plant health. SEM effectively handles high-dimensional soil data by enabling the simultaneous analysis of multiple observed and latent variables, managing multicollinearity, reducing data dimensionality, and providing tools to test complex hypotheses. In soil science, this makes SEM a powerful tool for understanding the intricate ways in which macronutrients influence soil fertility and plant nutrition, ultimately leading to more informed soil management and agricultural practices (Ren et al. 2024).

Why SEM Is Particularly Useful in Soil Science Research?

Structural Equation Modelling (SEM) is particularly useful in soil science research because it allows researchers to explore and quantify complex relationships between multiple variables, both observed and latent, in a way that traditional statistical methods cannot. Soil systems are inherently complex, involving numerous interacting factors, including soil physical properties, chemical properties, microbial communities, and overall soil quality. SEM's ability to model these interactions simultaneously makes it an invaluable tool in understanding and managing soil health (Sheikh et al. 2024).

One of the primary reasons SEM is useful in soil science is its capacity to handle latent variables, which are constructs that cannot be measured directly but are inferred from observed data. For example, soil quality is a latent variable that can be influenced by a combination of soil physical properties (like texture, structure, and bulk density), soil chemical properties (such as pH, nutrient levels, and cation exchange capacity), and biological factors (like microbial activity and biodiversity). Using SEM, soil scientists can create a model where soil quality is the latent variable explained by these various observed indicators. This provides a comprehensive picture of how different aspects of the soil contribute to its overall quality.

Another advantage of SEM in soil science is its ability to model complex causal relationships and indirect effects. For instance, suppose you are studying the effect of soil compaction (a physical property) on plant growth. Direct effects of compaction might include reduced root penetration and water infiltration. However, compaction could also have indirect effects by altering soil microbial communities, which in turn affect nutrient cycling and availability. SEM can model these direct and indirect pathways, allowing researchers to quantify the overall impact of compaction on plant growth, including both the immediate physical impacts and the secondary effects mediated through changes in soil biology.

SEM is also valuable in soil science because it can account for measurement errors in the observed variables. Soil properties are often measured with some degree of uncertainty due to variability in sampling, measurement techniques, and environmental conditions. SEM can incorporate these errors into the model, ensuring that the estimates of the relationships between variables are not biased by measurement inaccuracies. For example, if you're studying soil nutrient content and its effect on crop yield, SEM can adjust for the variability in nutrient measurements caused by differences in soil sampling depth or laboratory analysis methods, leading to more accurate conclusions (Liu et al. 2024).

In the study of soil chemical properties, SEM can be used to explore how different nutrients interact to influence plant growth. Soil nutrients like nitrogen (N), phosphorus (P), and potassium (K) often have synergistic or antagonistic effects. For instance, while nitrogen is crucial for plant growth, its availability can be affected by soil pH and organic matter content, both of which also influence other nutrients. SEM allows researchers to model these complex interdependencies and determine how changes in one nutrient might cascade through the system to affect plant health. This holistic approach is particularly useful for optimizing fertilization strategies and improving soil management practices.

SEM's ability to model hierarchical data structures is another reason it is particularly suited to soil science. Soil data often have a nested structure, such as samples taken from different depths within a single soil profile, or data collected from multiple plots within a field. SEM can handle these hierarchical relationships by allowing for multi-level models, where data from different levels are modeled simultaneously. For example, a soil scientist might use SEM to model the effects of soil depth on nutrient availability, taking into account how these effects vary between different plots or management practices. This allows for more nuanced analyses that can reveal patterns that might be missed by simpler statistical methods (Hu et al. 2024a).

In soil microbial ecology, SEM can be used to study the relationships between soil microbial communities and other soil properties. For instance, researchers might be interested in how soil pH, organic matter, and moisture content influence the diversity and activity of soil microbes. SEM allows for the inclusion of microbial diversity as a latent variable, influenced by these chemical and physical soil properties. The model can then be used to explore how changes in the microbial community, in turn, affect soil processes like nutrient cycling and decomposition. This can provide insights into the role of microbes in maintaining soil health and inform practices that promote beneficial microbial communities.

Additionally, SEM is particularly powerful in exploring the effects of land management practices on soil quality. For example, different tillage practices might affect soil structure, erosion rates, and organic matter content, all of which are interrelated. SEM can be used to model these relationships, allowing researchers to understand the combined effects of tillage on soil quality. By identifying the pathways through which management practices influence soil health, SEM can help guide the development of sustainable agriculture practices that optimize soil function while minimizing negative impacts. Finally, SEM's flexibility allows for the inclusion of feedback loops and reciprocal relationships, which are common in soil systems. For instance, soil organic matter can influence soil structure, which in turn affects water retention and erosion, ultimately feeding back into organic matter content through processes like decomposition and erosion deposition. SEM can model these feedback loops, providing a more realistic representation of soil processes and their long-term implications for soil health and sustainability (Chen et al. 2024).

SEM is particularly useful in soil science research because it can model complex, multi-faceted relationships between soil properties, account for latent variables like soil quality, handle hierarchical data structures, and explore both direct and indirect effects. Whether studying soil physical properties, chemical properties, microbial communities, or the impact of land management practices, SEM provides soil scientists with a powerful tool to untangle the complexities of soil systems and make informed decisions to improve soil health and agricultural productivity (Carrasco-Espinosa et al. 2024).

Practical Application in Soil Physical, Chemical, Biological, and Microbial Data

The practical application of Structural Equation Modelling (SEM) in soil science is transformative, enabling researchers to explore the intricate web of interactions among soil physical, chemical, biological, and microbial properties. By employing SEM, soil scientists can move beyond traditional correlation analyses, delving into the causal relationships that underpin soil processes and functions. For instance, understanding how soil texture, structure, and compaction affect water retention and infiltration requires a holistic approach, as these physical properties are not isolated but interconnected with chemical and biological aspects of the soil (Jarvis et al. 2024).

When applying SEM to soil physical properties, researchers can investigate how these factors influence the soil's ability to support plant growth. For example, soil compaction reduces pore space, impacting water infiltration and root penetration. This physical constraint can then be linked to chemical processes, such as the reduced movement of nutrients and oxygen in the soil, which in turn affects microbial activity. SEM allows for the modeling of these direct and indirect effects, providing a comprehensive picture of how physical properties like bulk density and porosity interact with other soil attributes to influence plant health (Liu et al. 2024).

In terms of soil chemical properties, SEM is particularly useful in understanding nutrient cycling and availability. Soil nutrients like nitrogen, phosphorus, and potassium do not act independently; their availability is influenced by soil pH, organic matter content, and microbial processes. SEM can model these relationships, showing how a change in one chemical property might cascade through the system to affect nutrient availability and plant uptake. For example, an increase in soil organic matter might enhance microbial activity, leading to greater nitrogen mineralization and improved plant nutrition. By applying SEM, researchers can quantify these pathways and identify the most influential factors in nutrient dynamics.

SEM also offers valuable insights into the biological aspects of soil, particularly in understanding how soil organisms contribute to ecosystem functions. Soil biological properties, such as the abundance and diversity of earthworms, insects, and other fauna, play critical roles in processes like decomposition and nutrient cycling. For example, earthworms influence soil structure through their burrowing activities, which can enhance water infiltration and root growth. SEM can be used to model these relationships, linking the presence of specific organisms to changes in soil physical and chemical properties, and ultimately to plant productivity. This approach allows for a better understanding of the roles that different organisms play in maintaining soil health (Na et al. 2024).

Microbial communities are another crucial component of soil systems, and SEM is particularly powerful in elucidating their roles. Soil microbes are responsible for a wide range of processes, including the decomposition of organic matter, nitrogen fixation, and the transformation of nutrients into forms that plants can use. SEM can model the relationships between microbial activity and other soil properties, such as how soil pH and moisture content influence microbial diversity and function. For instance, in acidic soils, SEM might reveal that reduced microbial diversity leads to lower rates of organic matter decomposition, which in turn affects nutrient availability for plants. By capturing these complex interactions, SEM provides a more nuanced understanding of how microbial communities contribute to soil fertility.

The practical application of SEM extends to exploring how land management practices impact soil health. Different farming practices, such as no-till, cover cropping, and organic amendments, can have varying effects on soil physical, chemical, biological, and microbial properties. SEM allows researchers to model these effects simultaneously, showing how changes in one aspect of the soil can influence others. For example, a no-till system might improve soil structure and organic matter content, which then enhances microbial activity and nutrient cycling. By using SEM, researchers can quantify these effects and assess the overall impact of different management practices on soil health.

One of the key strengths of SEM in soil science is its ability to handle large, complex datasets, which are common in studies involving multiple soil properties. Soil scientists often collect data on a wide range of variables, including physical, chemical, biological, and microbial properties, across different spatial and temporal scales. SEM can integrate these diverse data sources, allowing for the analysis of how these variables interact within the soil system. For instance, in a study of soil health across different land uses, SEM could be used to model how changes in land use affect soil compaction, organic matter content, microbial diversity, and nutrient cycling, providing a comprehensive understanding of how human activities influence soil function (Ren et al. 2024).

Another practical application of SEM is in the assessment of soil quality indicators. Soil quality is a complex concept that encompasses various physical, chemical, and biological properties. SEM can be used to develop and validate models that identify the key indicators of soil quality, such as organic carbon content, aggregate stability, and microbial biomass. By modeling the relationships between these indicators and soil functions like nutrient cycling and water retention, SEM helps researchers and land managers identify the most critical factors to monitor and manage for maintaining soil health.

SEM is also valuable in understanding the temporal dynamics of soil processes. Soil properties and processes often change over time in response to environmental factors like climate, vegetation cover, and human management. SEM can be used to model these temporal changes, showing how soil properties evolve and how these changes affect ecosystem functions. For example, SEM could be used to study the effects of seasonal changes in soil moisture on microbial activity and nutrient cycling, providing insights into how climate variability impacts soil health over time (Sterner et al. 2024).

In studies of soil remediation and restoration, SEM offers a powerful tool for evaluating the effectiveness of different interventions. Soil remediation often involves complex processes, such as the removal of contaminants, the addition of amendments, and the re-establishment of vegetation. SEM can be used to model how these interventions influence soil physical, chemical, biological, and microbial properties, and how these changes contribute to the overall recovery of soil functions. For instance, SEM could be applied to a study of contaminated soils to model how the addition of organic amendments improves soil structure, microbial activity, and nutrient availability, leading to successful plant establishment.

The ability of SEM to model feedback loops is particularly useful in soil science, where many processes are interdependent. For example, soil organic matter influences soil structure, which in turn affects water retention and erosion, and these physical changes can feedback to influence organic matter dynamics through processes like decomposition and erosion deposition. SEM allows for the modeling of these feedback loops, providing a more accurate representation of the complex interactions within the soil system. This capability is crucial for understanding the long-term sustainability of soil management practices and their effects on soil health.

Another practical application of SEM is in studying the impacts of climate change on soil systems. Climate change affects various soil properties, including temperature, moisture, and carbon cycling, all of which are interrelated. SEM can be used to model how changes in climate variables influence soil physical, chemical, biological, and microbial properties, and how these changes affect soil functions. For example, SEM might be used to explore how increased temperatures and altered precipitation patterns impact soil moisture, microbial activity, and organic matter decomposition, providing insights into how soil systems may respond to future climate scenarios.

In agricultural research, SEM is valuable for studying the interactions between soil properties and crop performance. Crop yields are influenced by a complex set of factors, including soil fertility, water availability, and microbial interactions. SEM can be used to model these relationships, helping researchers identify the key drivers of crop performance and develop strategies to optimize soil management for higher yields. For instance, SEM could be applied to a study of how soil nutrient levels, microbial diversity, and water retention collectively influence the growth and productivity of a particular crop, providing a holistic view of the factors that contribute to agricultural success.

Finally, the practical application of SEM in soil science is essential for informing policy and decision-making. Soil management policies often need to address multiple objectives, such as improving soil health, enhancing agricultural productivity, and protecting the environment. SEM provides a tool for evaluating the trade-offs and synergies between these objectives by modeling the complex relationships between soil properties and ecosystem services. For example, SEM could be used to assess how different land management practices impact soil carbon sequestration, water quality, and biodiversity, helping policymakers develop strategies that promote sustainable land use and soil conservation (Wang et al. 2024).

Applying SEM in Soil Science

Applying Structural Equation Modelling (SEM) in soil science provides a powerful framework for young scientists to explore and understand the complex interactions between various soil properties and plant nutrition. SEM allows researchers to move beyond simple correlations, enabling them to test hypotheses about the causal relationships that govern soil behavior and plant growth. For example, in sandy soils, which are often characterized by low fertility and poor water retention, SEM can help uncover the intricate dynamics between soil physical properties, chemical properties, and organic carbon pools (Xu et al. 2024).

When studying soil physical properties in sandy soils, SEM can be used to model how factors like soil texture and bulk density influence water retention and plant available water. Sandy soils typically have large pore spaces, leading to rapid drainage and low water-holding capacity. This physical limitation can directly impact plant growth by reducing the amount of water available for root uptake. SEM allows young scientists to quantify these relationships and explore how changes in soil texture or compaction might improve water retention. For instance, SEM could be used to assess the effectiveness of adding organic amendments, such as compost or biochar, in increasing the soil's ability to retain water and support plant growth.

Soil chemical properties are also crucial in determining the fertility of sandy soils. Nutrient availability in these soils is often limited due to the low cation exchange capacity (CEC) and poor retention of nutrients. SEM can model the relationships between soil chemical properties like pH, nutrient levels, and CEC, providing insights into how these factors interact to influence plant nutrition. For example, a young scientist might use SEM to investigate how the addition of lime (to raise pH) and organic matter (to increase CEC) could work together to enhance the availability of essential nutrients such as calcium, magnesium, and potassium. This type of analysis helps in identifying the most effective soil management practices for improving fertility in sandy soils.

One of the key applications of SEM in soil science is in understanding the role of soil organic carbon pools in maintaining soil health and fertility. Organic carbon in the soil is a critical component of soil structure, nutrient cycling, and water retention. In sandy soils, the organic carbon content is often low, making it a limiting factor for plant growth. SEM can be used to model how different soil management practices influence the size and stability of soil organic carbon pools. For instance, by applying SEM, a young scientist could explore how practices like crop rotation, cover cropping, or reduced tillage might increase organic carbon levels in sandy soils and thereby enhance soil fertility and resilience.

SEM is particularly useful in capturing the indirect effects that soil properties can have on plant nutrition. For example, increasing soil organic carbon not only improves soil structure and water retention but also enhances microbial activity, which plays a key role in nutrient cycling. By using SEM, young scientists can model these complex pathways and quantify how changes in one soil property might indirectly influence plant nutrition through multiple mechanisms. This holistic approach is essential for understanding the full impact of soil management practices on plant growth and for developing strategies that optimize soil health and productivity.

In sandy soils, where nutrient leaching is a significant concern, SEM can be used to study the interactions between soil physical and chemical properties that influence nutrient retention. Sandy soils are prone to losing nutrients like nitrogen and potassium through leaching, especially under heavy rainfall or irrigation. SEM can help young scientists model how factors such as soil texture, organic matter content, and pH interact to reduce nutrient leaching and improve nutrient use efficiency. For instance, the addition of organic amendments might not only increase soil carbon pools but also enhance nutrient retention by improving soil structure and CEC, leading to more sustainable nutrient management.

Another application of SEM in soil science is in evaluating the effectiveness of soil conservation practices in maintaining or improving soil health. For example, in sandy soils, conservation practices like mulching, cover cropping, and reduced tillage can help protect soil from erosion, enhance organic matter levels, and improve water retention. SEM allows researchers to model how these practices influence multiple soil properties simultaneously, providing a comprehensive assessment of their impact on soil health. This approach is particularly valuable for young scientists who are interested in developing sustainable soil management strategies that balance productivity with environmental protection.

In the context of plant nutrition, SEM can be used to explore how different soil properties interact to influence nutrient uptake and crop yield. For example, sandy soils often require careful management to ensure adequate nutrient availability throughout the growing season. By applying SEM, a young scientist could model the relationships between soil physical properties (such as texture and porosity), chemical properties (such as nutrient levels and pH), and plant nutrient uptake. This type of analysis can reveal how different soil management practices, like fertilization or irrigation, might need to be adjusted to optimize nutrient availability and improve crop performance in sandy soils.

SEM also provides a valuable tool for studying the long-term effects of soil management practices on soil organic carbon pools. In sandy soils, maintaining or increasing organic carbon levels is critical for sustaining soil fertility and resilience. SEM can be used to model how different practices, such as organic amendments, crop rotations, and reduced tillage, influence the size and stability of soil organic carbon pools over time. This type of analysis is important for young scientists who are interested in developing long-term soil management strategies that enhance soil carbon sequestration and mitigate climate change.

In addition to its applications in field studies, SEM can also be used in laboratory experiments to investigate the effects of specific soil treatments on soil properties and plant growth. For example, a young scientist might conduct a controlled experiment to test how different types of organic amendments affect soil physical properties, nutrient availability, and plant growth in sandy soils. By applying SEM to the experimental data, they can model the direct and indirect effects of the amendments on soil and plant outcomes, providing insights into the mechanisms that drive these effects and guiding the development of more effective soil management practices.

One of the strengths of SEM is its ability to integrate data from multiple sources, including field observations, laboratory experiments, and remote sensing. In the context of sandy soils, SEM can be used to combine data on soil physical properties, chemical properties, and organic carbon pools to develop a comprehensive model of soil health and fertility. For example, remote sensing data on soil moisture and vegetation cover can be integrated with field measurements of soil texture and nutrient levels to model how these factors influence soil carbon dynamics and plant productivity. This integrative approach is particularly valuable for young scientists who are interested in studying soil processes at multiple scales.

The application of SEM in soil science also extends to exploring the interactions between soil properties and environmental factors, such as climate and land use change. For example, in sandy soils, changes in temperature and precipitation patterns can have significant effects on soil moisture, nutrient cycling, and organic carbon levels. SEM can be used to model how these environmental changes interact with soil properties to influence plant growth and soil health. This type of analysis is important for young scientists who are interested in understanding the impacts of climate change on soil systems and developing strategies to mitigate these impacts. Applying SEM in soil science helps young scientists develop critical thinking and analytical skills. By working with complex models that integrate multiple soil properties and processes, they learn how to design experiments, collect and analyze data, and interpret results in a way that advances our understanding of soil systems. SEM encourages a holistic approach to soil science, where the focus is on understanding the interactions and feedback loops that drive soil behavior. This perspective is essential for developing sustainable soil management practices that support agricultural productivity, environmental health, and climate resilience (Xie et al. 2024).

Software for SEM

Several software packages are commonly used for conducting Structural Equation Modeling (SEM), each with its unique features and strengths. Below are some of the most widely used SEM software tools:

1. IBM SPSS Amos: SPSS Amos (Analysis of Moment Structures) is a user-friendly SEM software that integrates with IBM SPSS Statistics. It is widely used for its graphical interface, which allows users to draw models rather than writing code. Amos is suitable for those who prefer a visual approach to SEM and is often used in social sciences, psychology, and educational research. It supports a variety of SEM techniques, including confirmatory factor analysis (CFA), path analysis, and latent variable modeling.
2. Mplus: Mplus is one of the most powerful and flexible SEM software packages available. It supports a wide range of statistical analyses, including SEM, multilevel modeling, growth modeling, and Bayesian analysis. Mplus is known for its ability to handle complex models and large datasets, making it a popular choice among researchers in fields like psychology, education, and social sciences. The software uses syntax-based commands, which allows for precise control over model specification.
3. LISREL: LISREL (Linear Structural Relations) is one of the oldest SEM software tools and has been a foundational tool in the development of SEM techniques. It is particularly known for its robustness and ability to handle complex

models. LISREL is used primarily in academic research and offers both graphical and syntax-based interfaces. It supports a wide range of SEM methods, including CFA, path analysis, and structural regression models.

4. R (lavaan package): R is an open-source statistical software environment, and the lavaan package is a powerful tool for conducting SEM. lavaan is highly flexible, allowing users to specify complex models using R's scripting language. It is popular in academic and research settings due to its cost-effectiveness (free) and the extensive community support available. In addition to SEM, lavaan supports CFA, growth curve modeling, and multilevel SEM. The package is particularly attractive to those who are already familiar with R or those who need to perform SEM as part of broader statistical analyses.
5. SmartPLS: SmartPLS is a software package designed for Partial Least Squares Structural Equation Modeling (PLS-SEM), which is a variance-based approach to SEM. SmartPLS is particularly useful for exploratory research and when working with complex models that involve many variables. It is widely used in marketing research, business, and management studies. The software has a graphical user interface, making it accessible to users who prefer not to write code. SmartPLS is also known for handling smaller sample sizes compared to traditional SEM methods.
6. OpenMX: OpenMX is an open-source software for SEM that runs within the R environment. It is designed for advanced users who need to perform complex SEM analyses, including multilevel models, twin studies, and other genetic data analyses. OpenMX supports both frequentist and Bayesian approaches and is highly customizable. It is ideal for users who need to work with very large datasets or perform highly specialized analyses.
7. EQS: EQS is a SEM software package known for its robustness in handling non-normal data and complex model estimation. It includes advanced features like robust maximum likelihood estimation and methods for dealing with missing data. EQS is often used in social sciences, education, and behavioral research. The software offers both a graphical interface and syntax-based commands, catering to a wide range of users.
8. Stata (sem command): Stata is a comprehensive statistical software package that includes a command (sem) for performing SEM. Stata is widely used in economics, epidemiology, sociology, and political science. The SEM module in Stata supports a variety of models, including CFA, path analysis, and latent growth curve modeling. Stata's integration of SEM with its extensive data management and statistical analysis capabilities makes it a versatile tool for researchers.
9. SAS (PROC CALIS): SAS is a powerful statistical software suite that includes PROC CALIS, a procedure for conducting SEM. PROC CALIS is highly flexible and can handle a wide range of SEM models. SAS is often used in fields like biostatistics, epidemiology, and social sciences due to its ability to manage and analyze large datasets. The SEM capabilities in SAS are complemented by its comprehensive data handling and programming features.

These software tools provide a range of options for conducting SEM, from user-friendly interfaces to advanced scripting environments. The choice of software often depends on the complexity of the model, the size of the dataset, the user's familiarity with statistical software, and the specific requirements of the research project.

Example Application in Soil Science

Structural Equation Modeling (SEM) offers a nuanced approach to unraveling the complexities inherent in soil science, providing a way to understand the intricate relationships between various soil properties and their impacts on ecological processes. For instance, consider a study aimed at assessing the interplay between soil quality, plant growth, and microbial communities in agricultural systems. In such a scenario, SEM can help elucidate how soil physical properties like texture and bulk density influence the chemical properties, such as nutrient availability and pH, which in turn affect plant nutrient uptake and microbial activity. By modeling these interactions, SEM allows researchers to capture both direct and indirect effects, offering a comprehensive understanding of the factors that drive soil health and productivity.

In agricultural systems, soil fertility is a key factor that directly impacts crop yield and sustainability. SEM can be applied to explore how different management practices, such as fertilization, crop rotation, and tillage, influence soil nutrient dynamics and organic matter content. For example, SEM might reveal that the addition of organic fertilizers not only increases soil nutrient levels but also enhances soil structure, leading to improved water retention and microbial activity. These improvements in turn promote better plant growth and yield. By using SEM, researchers can quantify the relative importance of these pathways, guiding the development of management practices that optimize soil health and agricultural productivity.

Soil microbial communities play a critical role in nutrient cycling, decomposition, and overall soil health. SEM can be used to investigate the relationships between soil microbial diversity, soil organic matter, and nutrient availability. For instance, a researcher might use SEM to model how variations in soil organic carbon influence microbial community composition, and how these microbial changes, in turn, affect the cycling of nutrients such as nitrogen and phosphorus. The model could show that higher organic carbon levels lead to increased microbial diversity, which enhances nutrient mineralization and availability for plants. Such findings highlight the importance of maintaining organic carbon levels in soil to support robust microbial communities and sustainable nutrient cycling.

In the context of environmental stressors, SEM can be applied to study how soil properties and management practices influence soil resilience to disturbances such as drought or heavy rainfall. For example, SEM could be used to assess how soil texture and organic matter content affect soil water retention and erosion resistance. A model might reveal that soils with higher organic matter content are better able to retain moisture during drought conditions, reducing stress on plants and maintaining crop yields. Additionally, SEM could demonstrate that soils with a more stable structure, resulting from good management practices, are less prone to erosion during heavy rainfall. These insights are valuable for developing strategies to enhance soil resilience in the face of climate change.

In sandy soils, which are typically low in organic matter and nutrients, SEM can be particularly useful for understanding how different soil amendments and management practices can improve soil fertility and productivity. For example, a study might use SEM to explore how the addition of biochar influences soil organic carbon pools, nutrient retention, and microbial activity. The model could show that biochar application increases soil carbon content, which enhances nutrient retention and provides a more favorable environment for microbial activity. These changes, in turn, lead to improved plant growth and yield in sandy soils. SEM provides a way to quantify these complex interactions, helping researchers identify the most effective strategies for managing sandy soils.

Soil erosion is a major concern in many agricultural and natural systems, leading to the loss of topsoil, nutrients, and organic matter. SEM can be applied to study the factors that contribute to soil erosion and the effectiveness of different conservation practices. For instance, a researcher might use SEM to model how vegetation cover, soil texture, and slope influence erosion rates. The model could reveal that increased vegetation cover significantly reduces erosion by stabilizing the soil surface and improving water infiltration. Additionally, SEM could demonstrate that soils with finer textures and higher organic matter content are less prone to erosion, providing evidence for the importance of maintaining soil structure and organic matter levels in erosion-prone areas.

In studies of soil contamination, SEM can be used to assess how different soil properties influence the mobility and bioavailability of contaminants. For example, a study might use SEM to explore how soil pH, organic matter content, and clay content affect the retention and movement of heavy metals in the soil. The model could show that soils with higher clay and organic matter content are better able to retain heavy metals, reducing their mobility and bioavailability to plants. These findings can inform remediation strategies, suggesting that increasing soil organic matter or adjusting pH could help mitigate the risks associated with soil contamination.

SEM is also valuable in studies that aim to understand the impacts of land-use change on soil properties and ecosystem services. For instance, a researcher might use SEM to model how the conversion of forest to agricultural land affects soil organic carbon levels, nutrient cycling, and water retention. The model could reveal that deforestation leads to a significant decline in soil organic carbon and nutrient levels, which in turn reduces soil fertility and water-holding capacity. Such insights are crucial for developing land management practices that minimize the negative impacts of land-use change on soil health and ecosystem services.

Finally, SEM provides a framework for integrating data from multiple sources, such as field observations, laboratory experiments, and remote sensing, to build comprehensive models of soil systems. For example, a study might combine field measurements of soil properties with remote sensing data on vegetation cover and climate variables to model how changes in land use and climate affect soil organic carbon pools and nutrient cycling. SEM allows researchers to account for the complex interactions between these factors, providing a holistic understanding of soil dynamics and informing the development of sustainable land management practices. This integrative approach is particularly valuable for addressing the complex challenges facing soil science in the context of global environmental change.

Conclusion

Structural Equation Modeling (SEM) stands as a critical tool in soil science, offering a powerful and nuanced approach to analyzing the complex, interrelated factors that define soil ecosystems. Its ability to simultaneously account for multiple variables and pathways—whether they pertain to soil physical properties, chemical dynamics, biological processes, or microbial communities—makes SEM invaluable for advancing our understanding of soil health and its impact on broader ecological and agricultural systems. By integrating data across various scales and sources, SEM not only enhances the precision and depth of soil research but also guides the development of sustainable land management practices that are informed by a comprehensive view of soil-environment interactions. In a field where variables are deeply interconnected and where understanding these connections is essential for maintaining and improving soil fertility, productivity, and resilience, SEM emerges as an indispensable tool for both research and practical applications in soil science.

References

Carrasco-Espinosa K, Avitia M, Santini NS et al (2024) Nutrient contents and microbial communities as mediators of the effects of land-use in ecosystem functioning in alpine ecosystems from Central Mexico. J Soils Sediments. https://doi.org/10.1007/s11368-024-03804-y

Chen L, Dini-Andreote F, Liu H et al (2024) Integrating variation in bacterial-fungal co-occurrence network with soil carbon dynamics. J Appl Ecol 61:36–50. https://doi.org/10.1111/1365-2664.14535

da Silva JG, Chagas CA, dos Santos Souza TG et al (2024) Using structural equation modeling to assess the genotoxic and mutagenic effects of heavy metal contamination in the freshwater ecosystems: a study involving *Oreochromis niloticus* in an urban river. Sci Total Environ 913:e169529. https://doi.org/10.1016/j.scitotenv.2023.169529

Hu B, Xie M, Zhou Y et al (2024a) A high-resolution map of soil organic carbon in cropland of southern China. Catena 237:e107813. https://doi.org/10.1016/j.catena.2024.107813

Hu J, Baiser B, James RT, Reddy KR (2024b) An analysis of long-term everglades stormwater treatment areas performance using structural equation models. Ecol Eng 198:e107130. https://doi.org/10.1016/j.ecoleng.2023.107130

Jarvis N, Coucheney E, Lewan E et al (2024) Interactions between soil structure dynamics, hydrological processes, and organic matter cycling: a new soil-crop model. Eur J Soil Sci 75:e13455. https://doi.org/10.1111/ejss.13455

Le Geay M, Lauga B, Walcker R, Jassey VEJ (2024) A meta-analysis of peatland microbial diversity and function responses to climate change. Soil Biol Biochem 189:e109287. https://doi.org/10.1016/j.soilbio.2023.109287

Li J, Liu Y, Cui X et al (2024) Mycorrhizal mediation of soil carbon in permafrost regions depends on soil nutrient stoichiometry and physical protection. Sci Total Environ 920:e170907. https://doi.org/10.1016/j.scitotenv.2024.170907

Liu Y, Liu X, Fu Z, Zhang D, Liu L (2024) Soil temperature dominates forest spring phenology in China. Agric For Meteorol 355:110141. https://doi.org/10.1016/j.agrformet.2024.110141

Michalski R, Zaleski S (2024) Success factors in management of IT service projects: regression, confirmatory factor analysis, and structural equation models. Information 15:105. https://doi.org/10.3390/info15020105

Moshagen M, Bader M (2024) semPower: general power analysis for structural equation models. Behav Res 56:2901–2922. https://doi.org/10.3758/s13428-023-02254-7

Na M, Liu X, Tong Z et al (2024) Analysis of water quality influencing factors under multi-source data fusion based on PLS-SEM model: an example of East-Liao River in China. Sci Total Environ 907:168126. https://doi.org/10.1016/j.scitotenv.2023.168126

Ren Z, Han X, Feng H et al (2024) Long-term conservation tillage improves soil stoichiometry balance and crop productivity based on a 17-year experiment in a semi-arid area of northern China. Sci Total Environ 908:e168283. https://doi.org/10.1016/j.scitotenv.2023.168283

Sheikh AT, Chaudhary AK, Mufti S, Davies S, Rola-Rubzen MF (2024) Soil fertility in mixed crop-livestock farming systems of Punjab, Pakistan: the role of institutional factors and sustainable land management practices. Agric Syst 218:e103964. https://doi.org/10.1016/j.agsy.2024.103964

Sterner P, Pargent F, Deffner D, Goretzko D (2024) A causal framework for the comparability of latent variables. Struct Equ Model Multidiscip J. https://doi.org/10.1080/10705511.2024.2339396

Wang N, Li JM, Zhou YF (2024) Mechanism of action of marine ecological restoration on ecological, economic, and social benefits—an empirical analysis based on a structural equation model. Ocean Coast Manag 248:e106950. https://doi.org/10.1016/j.ocecoaman.2023.106950

Xie W, Duan E, Chung B, Allison SD (2024) Advancing quantitative models of soil microbiology, ecology, and biochemistry. In: Paul EA, Frey SD (eds) Soil microbiology, ecology and biochemistry, 5th edn. Elsevier, pp 473–492. https://doi.org/10.1016/B978-0-12-822941-5.00016-8

Xu B, Li J, Liu Y, Zhang T, Luo Z, Pei X (2024) Disentangling the response of vegetation dynamics to natural and anthropogenic drivers over the Qinghai-Tibet Plateau using dimensionality reduction and structural equation model. For Ecol Manag 554:e121677. https://doi.org/10.1016/j.foreco.2023.121677

Yang X, Song Z, Van Zwieten L et al (2024a) Significant accrual of soil organic carbon through long-term rice cultivation in paddy fields in China. Glob Chang Biol 30:e17213. https://doi.org/10.1111/gcb.17213

Yang X, Zheng R, Wan Z, Zhang Z (2024b) Differences of soil carbon pools and crop growth across different typical agricultural fields in China: the role of geochemistry and climate change. Environ Res 260:119623. https://doi.org/10.1016/j.envres.2024.119623

Yang Y, Chen Q, Zhou Y, Yu W, Shi Z (2024c) Soil bacterial community composition and function play roles in soil carbon balance in alpine timberline ecosystems. J Soils Sediments 24:323–336. https://doi.org/10.1007/s11368-023-03627-3

Zhao Y, Zhang M, Liu Z, Yang F (2024) Migration and transformation of soil phosphorus by organic acids: a global meta-analysis. J Soils Sediments 24:589–602. https://doi.org/10.1007/s11368-023-03665-x

Zhu Y, Liu C, Peng C et al (2024) Applications of structural equation modeling in plant functional trait research. Canadian Science Publishing. https://doi.org/10.1139/er-2023-0128

Index

T. Souza, *Advanced Statistical Analysis for Soil Scientists*,
https://doi.org/10.1007/978-3-031-88161-9

R

S

V

The manufacturer's authorised representative in the EU is Springer Nature Customer Service Centre GmbH, Europaplatz 3, 69115 Heidelberg, Germany. If you have any concerns regarding our products, please contact ProductSafety@springernature.com

Printed and bound by CPI Group (UK) Ltd, Croydon, CR0 4YY
01/10/2026
02225944-0012